Carbon–Carbon and Carbon–Proton NMR Couplings

Methods in Stereochemical Analysis

Volume 2

Series Editor

Alan P. Marchand

North Texas State University
Department of Chemistry
Denton, Texas 76203

Carbon–Carbon and Carbon–Proton NMR Couplings:

Applications to Organic Stereochemistry and Conformational Analysis

By

James L. Marshall

Verlag Chemie International

Deerfield Beach, Florida

CHEMISTRY

James L. Marshall, Ph.D.
Professor of Chemistry
North Texas State University
Denton, TX 76203

Library of Congress Cataloging in Publication Data

Marshall, James L., 1940-
    Carbon-carbon and carbon-proton NMR couplings.

    (Methods in stereochemical analysis; v. 2)
    Includes bibliographices and index.
    1. Nuclear magnetic resonance spectroscopy.
2. Proton magnetic resonance spectroscopy.   3. Stereo-
chemistry.   4. Conformational analysis.   I. Title.
II. Series.
QD272.S6M37   1983   547.3'0877   82-16117
ISBN 0-89573-113-4

Printed in the United States of America.

ISBN 0–89573–113–4 Verlag Chemie International, Deerfield Beach
ISBN 3–527–25977–5 Verlag Chemie GmbH, Weinheim

PREFACE

This monograph includes a comprehensive survey of long-range carbon–proton (^{13}C—^{1}H) and carbon–carbon (^{13}C—^{13}C) couplings. "Long-range coupling" describes any coupling greater than one bond. Early carbon–proton and carbon–carbon coupling studies dealt only with one-bonded couplings, whose larger values rendered them obvious and more easily procurable. More recently, the smaller long-range couplings have been obtained by the more sophisticated techniques available during the past 10 years.

These long-range couplings have a feature not inherent in one-bonded couplings. Whereas one-bonded couplings have no analogy with proton–proton (^{1}H—^{1}H) couplings in organic chemistry (the only model is the hydrogen molecule!), long-range couplings can be compared with geometrically equivalent proton–proton systems. This comparison leads to recognizable correlations between the carbon–proton/carbon–carbon couplings and the proton–proton couplings, which in turn allows conformational and stereochemical analysis. These long-range couplings can play a unique role in organic chemistry, and this monograph is devoted exclusively to this domain, with a minimum of discussion regarding one-bonded couplings. Readers interested in the topic of one-bonded couplings have access to other sources—NMR texts routinely include discussions of one-bonded carbon–proton couplings and reviews exist on one-bonded carbon–carbon couplings, notably one by Maciel (In "NMR Spectroscopy of Nuclei Other Than Protons," T. Axenrod and G. A. Webb, ed., Wiley-Interscience, New York, 1974). A convenient source discussing both one-bonded carbon–proton and carbon–carbon couplings is the excellent monograph by J. B. Stothers ("Carbon-13 NMR Spectroscopy," Academic Press, New York, 1972).

Also not included in this monograph are principles of nuclear magnetic resonance and Fourier transform techniques. However, special Fourier transform considerations are discussed that apply particularly to the acquisition of carbon coupling data. It is further assumed that the reader has a basic knowledge of proton–proton couplings, although continual reference is made to such couplings as they apply to carbon–proton and carbon–carbon couplings.

No specific discussion of chemical shifts is made, except as they apply to carbon couplings (for example, in the use of carbon couplings to make chemical shift assignments). However, the carbon chemical shifts of compounds affording carbon couplings are included in the tables if these chemical shifts have been reported simultaneously with the couplings. Where no carbon chemical shifts have been reported in the original publications, no attempt has been made to garner these shifts from other sources, because (if for no other reason) chemical shift assignments involving carbon couplings are more reliable.

Because of the low natural abundance of the carbon-13 isotope, the determination of signs of carbon couplings is particularly troublesome. Special consideration is given to this problem in Chapter 5 (Special Topics). Where no J values are listed in the tables, this does not imply that the coupling is zero (although it often is). Instead, the absence of a J value means simply that the value was not available in the original publication, whether because the coupling was not measured, was unavailable because of broadening of the peaks, or was not procurable owing to the presence of chemically equivalent nuclei.

CONTENTS

1

INTRODUCTION

Hydrogen (^{1}H) nuclear magnetic resonance (NMR) has been the most powerful method of conformational analysis in the past 20 years. The success of ^{1}H NMR has hinged upon several features, principally the higher natural abundance (99.98%), the high sensitivity of ^{1}H, and its spin of $\frac{1}{2}$ (leading to simpler spectra). The near ubiquitousness of hydrogen in organic molecules also makes it a ready handle in the analysis of organic compounds.

There are some situations, however, when ^{1}H NMR is not useful. These situations occur when there are not enough hydrogens (as in halogenated organics), when there are too many hydrogens (leading to spectra too complex to analyze), or when the hydrogens are in the wrong places (e.g., when they are too far apart in the molecule to be mutually coupled). In such instances it would be desirable to have an additional tool for the analysis of the geometry of molecules, and attention turns naturally to an inspection of other potentially suitable nuclei in NMR analysis; Table 1-1 lists likely candidates. In this table, only nuclei with spins of $\frac{1}{2}$ are included, because of the more complex spectra and because of quadrupolar broadening, which occurs with nuclei with spins of 1 and above.

From Table 1-1 the most obvious candidate is fluorine (^{19}F), with an isotopic abundance of 100% and a sensitivity almost that of ^{1}H. However, ^{19}F is a comparatively scarce element in compounds. Additionally, electronegativity effects and indirect contributions arising from nonbonded electrons do not render analysis as straightforward as for ^{1}H. Therefore, ^{19}F does not lend itself so readily to conformational analysis, and ^{19}F couplings accordingly have been predominantly involved in theoretical considerations.

The next most likely candidate in Table 1-1 might be phosphorus (^{31}P). This nucleus has a high natural abundance (100%) and a high sensitivity, and indeed ^{31}P spin–spin coupling has been used to a large extent in conformational analysis. Apparently the complications that arise from the nonbonded electrons in fluorine do not occur here, and dihedral angle relationships have been set up that appear to work well. The only drawback to ^{31}P is that phosphorus NMR is limited to the relatively small number of compounds that bear this nucleus.

TABLE 1-1. SMALLER NUCLEI WITH $I = \frac{1}{2}$ — NATURAL ABUNDANCES
AND SENSITIVITIES[4]

Nucleus	Natural abundance (%)	Natural sensitivity for equal number of nuclei
^1H	99.98	1.00
^{13}C	1.11	0.0159
$^{15}N^a$	0.37	0.00104
^{19}F	100.0	0.833
$^{29}Si^a$	4.70	0.00784
^{31}P	100.0	0.0663

^{a} ^{15}N and ^{29}Si have spins of $-\frac{1}{2}$.

Accordingly, organophosphorus compounds of biochemical interest have been extensively studied.

Remaining in Table 1-1 are nuclei of lower natural abundance. Nuclear magnetic resonance analysis of compounds containing these nuclei would suffer from decreased signal strength. Nevertheless, advanced time-averaging techniques, notably Fourier transform NMR spectroscopy, have allowed ready analysis of all of the nuclei appearing in the table at natural abundance levels. Furthermore, in the past 10 years or so, enriched isotopic material has become available at reasonable prices for all these nuclei (except ^{29}Si), allowing enhanced signal to noise ratios for those who choose the additional effort of synthetically incorporating the isotope into a compound.

The most attractive of these lower abundance nuclei is perhaps ^{13}C: Its natural abundance is reasonable (ca. 1%), its sensitivity is about as great as any NMR-active nucleus, and enriched ^{13}C material for synthetic incorporation is readily available (about \$800 for a mole of 90% $^{13}CO_2$). Furthermore, the absence of nonbonding electrons in carbon can be expected to minimize unexpected mechanistic contributions to couplings. In fact, a natural question is: with all of the rich background available in ^1H NMR, and because carbon is the most similar electronically to hydrogen in the covalent state (i.e., no nonbonding electrons), is it possible that the geometrical dependence of carbon couplings is similar to that of proton couplings? More specifically, can proton couplings be used as models to anticipate carbon couplings? If so, then conformational analysis using carbon would be facile; and in compounds where proton couplings cannot be used, carbon couplings may prove to be a valuable addition to the arsenal already available to chemists interested in conformational analysis and stereochemistry of carbon compounds.

Practical Considerations: Obtaining the Data

Carbon–proton couplings may be extracted from the proton NMR spectra of natural abundance compounds in special cases, viz., a simpler system where the main proton signals do not mask the much weaker signals arising from compounds containing the natural abundance ^{13}C isotope. In practice, only large carbon–proton splittings may be observed, because the smaller splittings are not spread out sufficiently to emerge from under the intense center band of the normal proton signals. Generally only those carbon–proton couplings that are directly bonded ($^{1}J_{CH}$) are large enough (120–250 Hz) to be directly observable. These patterns, attributable to the directly bonded carbon–proton couplings, flank the more intense center bands as "satellite spectra." Unfortunately, these satellite spectra can be quite complex and only the simplest molecules are amenable to satellite spectral analysis.

A different approach to obtaining carbon–proton couplings is to observe these splittings for natural abundance compounds in their ^{13}C NMR spectra. This approach suffers from two difficulties. First, for a compound with x different carbon atoms, there appear x different spectra, all appearing simultaneously and potentially overlapping in an indiscernable manner. Second, the sensitivity of the carbon-13 nucleus is low, as noted in Table 1-1, and obtaining satisfactory spectra can be laborious. Nevertheless, a limited amount of carbon–proton coupling data have been obtained by this method in suitably chosen systems. An example is shown in Figure 1-1, which shows the proton-coupled ^{13}C spectrum of bromobenzene analyzed to obtain all C—H coupling constants.[1]

A further area of exploration involving carbon-13 is carbon–carbon couplings. An experimental benefit here is the possibility of using proton decoupling. Proton decoupling helps alleviate the two problems noted immediately above: First, because carbon–proton coupling is removed, simplified spectral patterns result; second, proton decoupling generally enhances carbon signals, owing to the nuclear Overhauser effect. Unfortunately, because of the low natural abundance of carbon-13, the chance of two carbon-13 nuclei existing simultaneously at two given positions in a molecule is quite small. Time averaging can enhance these carbon–carbon satellite signals so that they are measurable if the values are sufficiently large to be apart from the large center band. With a few exceptions, only directly bonded carbon–carbon coupling constants ($^{1}J_{CC}$, typically 30–100 Hz) can be obtained by this method.

In view of the difficulties in studies involving natural abundance carbon-13, using enriched samples is an obvious advantage. In the past 10–15 years (as noted above), carbon-13 precursors have become available at reasonable prices, allowing the economically reasonable synthesis of molecules with a specific carbon-13 label. For example, both 90% $^{13}CO_2$ and 90% ^{13}CO (carbon-13 labeled carbon dioxide and carbon monoxide, respectively) are available at ca. $800 a mole, and 90% $^{13}CH_3I$ (carbon-13 labeled methyl iodide) is available at

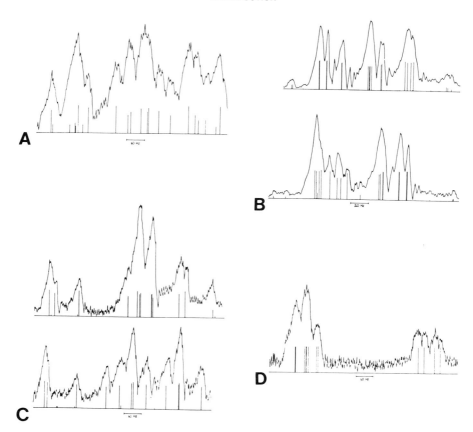

Figure 1-1. Natural abundance proton-coupled ^{13}C spectrum of bromobenzene. (A) A portion of the observed ^{13}C NMR spectrum of C_s in bromobenzene with calculated line plot. Time averaged for 512 scans at 0.2 Hz/cm (12 Hz sweep width) and 0.3 Hz/s. (B) Observed upfield (upper) and downfield (lower) ^{13}C NMR spectra of C_m in bromobenzene with calculated line plots. Time averaged at 0.4 Hz/cm and 0.6 Hz/s. (C) Observed portions of the upfield (upper) and downfield (lower) ^{13}C NMR spectra of C_o in bromobenzene with calculated line plots. Time averaged for 256 scans at 0.2 Hz/cm and 0.3 Hz/s. (D) A portion of the observed upfield ^{13}C NMR spectrum of C_p in bromobenzene with calculated line plot. Time averaged at 0.2 Hz/cm and 0.15 Hz/s. (Reproduced with permission.[1])

slightly higher price or may be synthesized in a two-step sequence from $^{13}CO_2$. Labeled potassium cyanide, $K^{13}CN$, is also commercially available but is much more expensive, and its synthesis in the laboratory is difficult. Although carbon-13 precursors of lower isotopic purity are available at lower prices, it is usually desirable to use isotopic enrichment above 90% so that the center-band problem does not obscure small couplings. Precursors with isotopic enrichment above 95% are commercially available, but the extra cost is not warranted. More complicated molecules, from acetic acid to steroid derivatives, singly labeled or multiply labeled, are commercially available from several sources or may be synthesized in any modestly equipped organic laboratory that includes a vacuum rack. As an example of the type of spectrum that may be obtained from carbon-13 labeled compounds, Figure 1-2 shows the ^{13}C NMR spectra of both natural and methyl ^{13}C-labeled o-nitrotoluene; in the labeled material the splittings caused by the carbon–carbon couplings are clearly apparent.[2]

With a specific position labeled with carbon-13, the proton spectrum would involve exclusively the proton system coupled with the labeled carbon. Carbon–carbon couplings could be obtained from these specifically labeled systems by observing the naturally occurring carbons, each of which potentially could be coupling with the labeled carbon. Thus, the naturally occurring carbon signals would each be split into a doublet, reflecting the coupling with the labeled carbon (Figure 1-2). With state of the art instrumentation available in the past few years, couplings as small as 0.5 Hz have been measured reproducibly with an accuracy of 0.1 Hz with electromagnet NMR spectrometers. Very recent superconducting NMR spectrometers can resolve routinely couplings as small as 0.1 Hz (if the T_2 value for the carbon in question is not exceptionally small; in these exceptional cases, the small T_2 value results in a broadened signal).

Synthesis of Carbon-13 Labeled Compounds

The synthesis of the labeled compounds most commonly begins by reacting a Grignard reagent with labeled carbon dioxide to form the ^{13}C-carboxyl-carboxylic acid (reaction 1). When a potentially stable carbonium ion is an intermediate, then a halide precursor can be reacted with labeled carbon monoxide with acid catalyst to form the carboxylic acid (reaction 2). An example of reaction 2 would be R = 1-adamantyl. Labeled carbon dioxide can also be carried to methyl Grignard (reaction 3), which can add in the usual manner to carbonyl compounds or can react in nucleophilic substitution reactions. Additionally, Wittig reagents (reaction 4) can be synthesized from the intermediate methyl iodide and can be carried on to labeled methylene compounds. Finally, it should be noted that labeled potassium cyanide is also available for synthesis of carbon-13 labeled compounds, but as mentioned above this reagent is not as available as ^{13}CO$_2$ and ^{13}CO. Reactions 1–4 open the way for a great variety of further reactions and derivatization limited only by imagination. In the literature these labeled compounds have been carried on to alkanes, olefins, acetylenes; carboxylic acids, aldehydes, ketones, nitriles, alcohols, and halides; acyclic and polycyclic compounds; aromatics and heterocyclic compounds; and

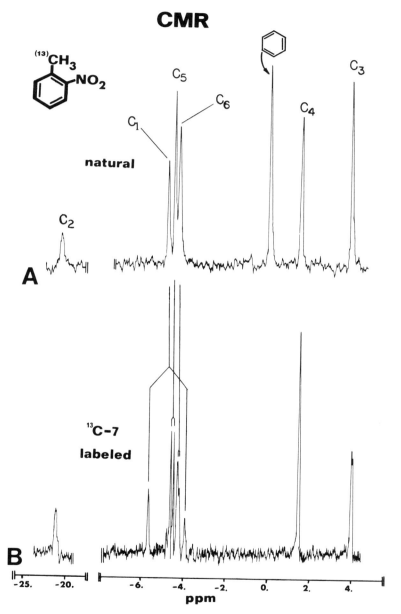

Figure 1-2. Carbon magnetic resonance spectra of (A) natural *o*-nitrotoluene and (B) ¹³C-7 labeled *o*-nitrotoluene. (Reproduced with permission.²)

compounds labeled internally (such as in an aromatic ring) or externally (as in a substituent). Experimental details are given below[3] for specific compounds illustrating the use of reactions 1–4. The synthesis of ^{13}C labeled acetic acid is a special case, owing to its solubility in water and the consequent difficulty in its isolation. This example for acetic acid is for doubly labeled compound to show that reactions 1 and 2 can be combined. For more detailed synthetic possibilities, the reader is referred to a recent treatise on the synthesis of compounds with specific enrichment with stable isotopes.[5]

Reaction 1: $\text{R—X} + \text{Mg} \longrightarrow \text{R—MgX} \xrightarrow[\text{(2) H}_3\text{O}^+]{\text{(1) }^{13}\text{CO}_2} \text{R—}^{13}\text{COOH}$

Reaction 2: $\text{R—X} \xrightarrow[\text{(2) H}_2\text{O}]{\text{(1) }^{13}\text{CO, H}_2\text{SO}_4} \text{R—}^{13}\text{COOH}$

Reaction 3: $^{13}\text{CO}_2 \xrightarrow{\text{LiAlH}_4} {}^{13}\text{CH}_3\text{OH} \xrightarrow{\text{HI}} {}^{13}\text{CH}_3\text{I} \xrightarrow{\text{Mg}} {}^{13}\text{CH}_3\text{MgI}$

Reaction 4: $^{13}\text{CH}_3\text{I} + \text{Ph}_3\text{P} \longrightarrow \text{Ph}_3\overset{+}{\text{P}}\text{—}^{13}\text{Me I}^- \xrightarrow{\text{BuLi}} \text{Ph}_3\text{P}=^{13}\text{CH}_2$

$$\begin{array}{c} \text{R} \\ \diagdown \\ \quad \quad \text{C}=\text{O} \\ \diagup \\ \text{R} \end{array}$$

$$\begin{array}{c} \text{R} \\ \diagdown \\ \quad \quad \text{C}=^{13}\text{CH}_2 \\ \diagup \\ \text{R} \end{array}$$

Example of Reaction 1: Preparation of Cyclohexanecarboxylic Acid-carboxyl-^{13}C *Acid.* A solution of cyclohexylmagnesium bromide was prepared under an argon atmosphere in a 250-ml Erlenmeyer flask with a 24/40 joint adaptable to a high-vacuum system, using 10.0 g of cyclohexyl bromide (0.061 mol), 1.51 g of magnesium turnings (0.062 mol), and 150 ml of anhydrous ether. A magnetic stirrer bar was placed in the reaction flask and the flask was connected to the vacuum system. The Grignard solution was degassed twice at 10^{-5} Torr by the liquid nitrogen freeze–thaw technique. The Grignard solution was then sealed from the vacuum system by closing the appropriate stopcock.

The volume of the vacuum system had been determined previously so that the weight of carbon dioxide introduced to the system could be determined by measuring the pressure of the manometer. One gram (0.0223 mol) of $^{13}\text{CO}_2$ was delivered to the vacuum system and then was condensed into the reaction vessel cooled with liquid nitrogen. The reaction vessel was then sealed from the vacuum system and allowed to warm by substituting a Dry Ice–isopropyl alcohol bath for the liquid nitrogen. After about 20 min the Grignard solution thawed sufficiently to permit stirring. The reaction mixture was then allowed to warm to 0° with stirring over a period of 1 h. The system was opened and water was added dropwise to the reaction mixture to react excess Grignard reagent. The reaction mixture was dissolved in 6 N hydrochloric acid and extracted with four 50-ml portions of ether. The ether was evaporated to give 2.52 g (86% based on the carbon dioxide).

Example of Reaction 2: Preparation of 1-Adamantanecarboxylic-carboxyl-^{13}C *Acid.* A mixture of 1.95 g of 1-bromoadamantane, 10 ml of cyclohexane, and 80 ml of concentrated sulfuric acid was prepared and introduced into a 100-ml round-bottom flask. This flask was then connected to the vacuum rack described above, to which 0.19 g of ^{13}CO (0.00909 mol) had been introduced. The contents of the flask were allowed to stir and react with the gaseous carbon monoxide at room temperature for 40 h (carbon monoxide does not condense even at liquid nitrogen temperatures). The flask was then removed (caution: unreacted carbon monoxide is poisonous) and was poured over 500 ml of ice. The resulting mixture was extracted with three 50-ml portions of ether. The ethereal extracts were then extracted with three 50-ml portions of 6 N aqueous sodium hydroxide solution. The aqueous extracts were reacidified with concentrated sulfuric acid (caution: heat generated), whereupon a white precipitate occurred. This precipitate was extracted out with three 50-ml portions of ether. The ether extract was washed with 30 ml of water and 30 ml of brine, dried (anhydrous magnesium sulfate), and concentrated to give 0.795 g (48%), m.p. 172.0–173.5°C. The yield of final product can be increased to 95% if the amount of carbon monoxide is doubled, but of course labeled carbon monoxide is lost.

Example of Reaction 3. Preparation of ^{13}C-Methyl Iodide. Lithium aluminum hydride (2.6 g, 0.069 mol) was suspended in 25 ml of bis(2-ethoxyethyl) ether in a flask suitable for the vacuum line. (This solvent must be free of peroxides before the lithium aluminum hydride is added. Failure to have peroxide-free solvent can result in a severe explosion.) After attachment to the vacuum line and freezing and evacuating the flask, labeled carbon dioxide (2.0 g, 0.045 mol) was condensed into the flask. The apparatus was allowed to warm slowly to room temperature. As the solvent thawed the flask filled with white smoke, which disappeared after a short period of stirring. The solution was stirred an additional 30 min at room temperature and then removed from the vacuum system. To the solution was carefully added 75 ml of benzyl alcohol and the solution was distilled. The methanol and water distilled first at less than 100°C. The mixture of methanol and water from above was added to 60 ml of 55% hydroiodic acid and carefully distilled. The methyl iodide distilled as it formed into a receiver cooled in a Dry Ice–acetone bath. The water that codistilled was decanted and the product dried over Drierite. The yield from carbon dioxide is nearly quantitative to give 6.3 g of methyl iodide.

Example of Reaction 4: Preparation of 2-^{13}C-Methylenebornane via a Wittig Reagent. A 250-ml three-neck flask was charged with sodium hydride (0.05 mol as a 55% dispersion in mineral oil) and then was washed with several portions of pentane to free it from the mineral oil while the flask was constantly purged with dry argon to exclude moisture and oxygen. The flask was then stoppered and fitted with a condenser and 30 ml of dimethyl sulfoxide was added and the mixture heated to 50° for 1 h until the evolution of hydrogen had ceased. The flask was then cooled in a ice bath and 0.05 mol (20.2 g) of ^{13}C-methyltriphenyl-phosphonium iodide [prepared from 0.05 mol (7.1 g) of ^{13}C-methyl iodide and 0.05 mol (13.1 g) of triphenylphosphine] was added as a solution in 50 ml of warm dimethyl sulfoxide. The resulting dark red solution of the ylid was warmed to room temperature and stirred for 10 min. To this solution, while it was stirring, 6.0 g (0.04 mol) of (+)-camphor were added as a solution in 10 ml of dimethyl sulfoxide. The reaction was then heated to 50°C and allowed to react overnight. Then the mixture was cooled and poured into 300 ml of pentane. The mixture was then transfered to a separatory funnel and 100 ml of water were added and the layers separated. The pentane layer was washed with 4 × 50 ml portions of water and these washes were combined and washed with 4 × 50 ml portions of pentane and these pentane washes were combined and added to the other pentane layer. The combined organic fractions were then washed with 6 × 100 ml portions of brine to free them from the dimethyl sulfoxide. The pentane layer was dried with magnesium sulfate and the solvent removed to give a thick oil. This was further purified by column chromatography using a 20 cm × 2 cm column of F-20 alumina eluting with pentane. The solvent was removed to give 4.5 g (75%) of a clear colorless oil, which solidified upon standing to give a crystalline mass, m.p. 65°.

Special Case of Reactions 1 and 2: Preparation of Potassium Acetate-1,2-[13]C_2. This was prepared by the normal Grignard procedure described in reaction 1 using ^{13}C-methyl iodide and ^{13}C-carbon dioxide, except that the acidified Grignard complex was steam distilled. The azeotrope containing dilabeled acetic acid was neutralized with a potassium hydroxide solution. The resulting potassium acetate solution was evaporated and then dried under high vacuum to give 3.8 g (0.038 mol) of product (82.5% yield).

Additional Techniques for Obtaining Carbon–Carbon Couplings

A modification of the techniques discussed above for studying carbon-13 labeled compounds is the use of two-dimensional Fourier transform (FT) techniques.[6] In two-dimensional FT spectroscopy, a T_2-like pulse sequence (90°–τ–180°) is performed for a variety of τ values, giving several spectral traces for each of the τ values selected. An additional Fourier transform of a specific frequency (resonance) gives the carbon–carbon coupling of that carbon to the labeled carbon, because the intensity (and phase) of the signal in question is modulated in accordance with the coupling constant. Because this method does not depend on field homogeneity, quite small J values (less than 1 Hz) can be measured accurately, and several of the longer range couplings ($^4J_{CC}$) of Table 3-2 were obtained by this method.

In a more recent technique,[7,8] carbon–carbon couplings have been procured where *no* labeled carbon is present. Instead, the carbon–carbon coupling arising from natural abundance carbons in measured by a "double-quantum coherence" experiment.[7,8] In this experiment, a unique pulse sequence is utilized $[90°(X)-\tau-180°(\pm Y)-\tau-90°(X)-\Delta-90°(\Phi)-\text{Acquisition}(\Psi)]$ that cancels the signal arising from molecules with one ^{13}C nucleus but that reveals the weak signals from molecules with two coupled carbon-13 spins. Simple systems are obvious candidates for this method, because all carbon–carbon couplings are simultaneously present in one spectrum. Examples studied by this method include tetramethyladamantane[7] and piperidine[8] (**178** and **179**, Tables 3-2). In this method, two-dimensional FT additionally can be used to obtain small couplings.[7] One drawback to this method is that coupling assignments can be ambiguous. Specifically, the original assignments for ^2J and ^4J in tetramethyladamantane (Table 3-2, **178a**) should clearly be reversed in view of the corresponding ^2J and ^4J values for 1-methyladamantane (Table 3-2, **21**), which are secure.[9]

References

1. Tarpley, A. R. Jr.; Goldstein, J. H. *J. Phys. Chem.*, **76**, 515 (1972).
2. Marshall, J. L.; Ihrig, A. M. *Org. Mag. Resonance*, **5**, 235 (1973).
3. Marshall, J. L.; Barfield, M. Unpublished results.
4. Becker, E. D. "High Resolution NMR"; Academic Press: New York, 1969, p. 241.
5. Ott, D. G. "Synthesis with Stable Isotopes"; Wiley-Interscience: New York, 1980.
6. Niedermeyer, R.; Freeman, R. J. *J. Mag. Resonance*, **30**, 617 (1978).
7. Bax, A.; Freeman, R.; Kempsell, S. P. *J. Mag. Resonance*, **41**, 349 (1980).
8. Bax, A.; Freeman, R.; Kempsell, S. P. *J. Am. Chem. Soc.*, **102**, 4849 (1980).
9. For an example using the double-quantum coherence method in conformational analysis, see Chapter 3, reference 38.

2

CARBON–PROTON COUPLINGS

Carbon–Proton Coupling Constants as a Test

A natural step before testing carbon–carbon couplings (J_{CC}) would be to examine the corresponding carbon–proton couplings (J_{CH}), for two reasons. First, if non-Fermi contact mechanisms[1] are operating for carbon–carbon couplings, this complication may be isolated by first observing carbon–proton couplings (and, in fact, non-Fermi contact mechanisms should be zero if one of the nuclei is a proton). Second, the hypothesis that carbon couplings are like proton couplings should be approached gingerly, lest other mechanistic complications arise. Accordingly, the approach in a proton–proton coupling system would be to substitute a carbon for one of the protons and observe the resulting carbon–proton coupling. Thereby, data would be accumulated to ascertain whether a correlation occurs between geometrically equivalent proton–proton couplings and carbon–proton couplings.

The Correlation of Carbon–Proton and Proton–Proton Couplings

In Table 2-1 are compiled the J_{CH} values, including signs, of the major types of couplings in ^{13}C-carboxyl-labeled carboxylic acids and these couplings are compared with geometrically equivalent proton–proton values.[2] Carboxylic acids were used because of the ease of synthesis (see Chapter 1). The carbon–proton couplings were obtained by means of the ^{1}H NMR spectrum via LAOCOON analysis[3] and spin-tickling techniques.[2,4] For each of the carbon–proton couplings, the proton–proton model used was simply that in which the carboxyl group had been substituted by a proton. Accordingly, the model for isocrotonic acid (1) is propene (11); and from compound 1 the $^2J_{CH}$, $^3J_{CH}$, and $^4J_{CH}$ values are compared with the geometrically equivalent $^2J_{HH}$, $^3J_{HH}$, $^4J_{HH}$ values of compound 11. A comparison of all of the J_{CH} values of

TABLE 2-1. J_{CH} VALUES OF ^{13}C-*carboxyl*-CARBOXYLIC ACIDS COMPARED WITH GEOMETRICALLY EQUIVALENT J_{HH} VALUES OF MODEL COMPOUNDS[a]

Carbon–proton systems			Proton–proton systems	
Compound	J_{HC}	Type of J	J_{HH}	Compound
HOOC, C=C, CH₃, H, H **1**	+3.12 +14.50 −1.28	^2J ^3J ^4J	+2.17 +16.98 −1.77	H, C=C, CH₃, H, H **11**
HOOC, C=C, H, H, CH₃ **2**	+3.39 +6.78 −0.85	^2J ^3J ^4J	+2.17 +10.09 −1.41	H, C=C, H, H, CH₃ **12**
HOOC, C=C, H, H, H **3**	+4.1 +7.6 +14.1	^2J ^3J(cis) ^3J(trans)	+2.5 +11.6 +19.1	H, C=C, H, H, H **12**
COOH (benzene ring) **4**	+4.08 +1.11 +0.48	^{3}(ortho) ^{4}(meta) ^{5}(para)	+7.7 +1.4 +0.6	H (benzene ring) **13**
CH₃—C≡C—COOMe **5**	−1.48	^4J	−2.93	CH₃—C≡C—H **14**
H—C≡C—COOH **6**	+4.5	^3J	+9.53	H—C≡C—H **15**

TABLE 2-1. (continued)

Carbon–proton systems			Proton–proton systems	
Compound	J_{HC}	Type of J	J_{HH}	Compound
7	-6.35 $+4.54$ $+2.51$	^2J $^3J(cis)$ $^3J(trans)$	-12.4 $+9.10$ $+3.70$	16
8	-6.17	^2J	-12.6	17
9	-4.61	^2J	-10.2	18
10	-6.89	^2J	-12.5	19

[a]Reproduced with permission from Wiley Heydon Ltd.[2]

Table 2-1 with the respective J_{HH} values of the model compounds shows that in general the J_{CH} values are roughly 0.6 the respective J_{HH} values. These couplings in Table 2-1 include geminal (^2J) olefinic and aliphatics, vicinal (^3J) acetylenic, olefinic (both cis and trans), aromatic and aliphatic (of differing geometries), and four-bonded (^4J) acetylenic and aromatic. A plot of the J_{CH} values vs. the respective J_{HH} values (see Figure 2-1) demonstrates reasonable correlation (correlation coefficient $= 0.975$) with a slope of 0.62.

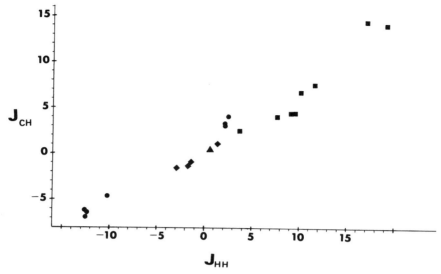

Figure 2-1. J_{CH} vs. J_{HH} for geometrically equivalent systems, as taken from Table 2-1. ● $= {}^2J$; ■ $= {}^3J$; ◆ $= {}^4J$; ▲ $= {}^5J$. (Reproduced with permission from Wiley Heydon Ltd.[2])

The data of Figure 2-1 have several features that may be of general use to conformational analysis. First, the cis and trans aliphatic couplings of compounds **7** and **16** indicate that dihedral angle may be a reliable indicator of carbon–proton couplings, just as the Karplus dihedral angular dependence[5] is for proton–proton couplings. Second, comparison of the cis and trans olefinic couplings in compounds **1, 2, 3, 11,** and **12** show that cis–trans geometries in olefins can be established by the use of carbon–proton couplings. Third, the well-behaved nature of the ortho, meta, and para couplings in the aromatic compounds **4** and **13** demonstrate that relative orientations in aromatic compounds may be determined by J_{CH} values. Fourth, the agreement of coupling signs is without exception for respective J_{CH} and J_{HH} values. The worst agreement is for the two-bonded (^2J) olefinic values (compounds **1, 2, 3, 11,** and **12**), but olefinic geminal couplings should be extremely sensitive to changes in geometry because the nodal plane of each of the orbitals involved should lie close to one of the coupling nuclei.[6,7]

To emphasize the significance of the correlation of Figure 2-1, a similar comparison can be made between ^1H and another nucleus that does have nonbonded electrons. Sufficient data exist for ^{19}F to tabulate such a comparison (Table 2-2). Study of this table shows that J_{HF} and J_{HH} do not correlate well; signs are not the same for all cases and absolute values are in very poor agreement (the correlation coefficient is -0.098).

That J_{CH}/J_{HH} from Figure 2-1 is 0.62 clearly illustrates that merely a

TABLE 2-2. A COMPARISON OF J_{HF} VALUES AND GEOMETRICALLY EQUIVALENT J_{HH} VALUES[a]

Proton–fluorine systems			Proton–proton systems		
Compound	J_{HF}	Type of J	J_{HH}	Compound	
cis-FHC=CHCH$_3$	+89.9	^2J	+2.17	H$_2$C=CHCH$_3$	
	+41.8	^3J	+16.98		
	+2.6	^4J	−1.77		
trans-FHC=CHCH$_3$	+84.8	^2J	+2.17	H$_2$C=CHCH$_3$	
	+19.9	^3J	+10.09		
	+3.3	^4J	−1.41		
FHC=CH$_2$	+84.7	^2J	+2.5	H$_2$C=CH$_2$	
	+20.1	^3J(cis)	+11.6		
	+52.4	^3J(trans)	+19.1		
CH$_3$—⟨ ⟩—F	+8.7	^2J	+7.7	⟨ ⟩—H	
	+5.8	^3J	+1.4		
FC≡CH	21	^3J	+9.53	HC≡CH	
(chlorinated bicyclic structure with F)	+55.1	^2J	−12.4	(chlorinated bicyclic structure)	
	+24.7	^3J(cis)	+9.10		
	+12.2	^3J(trans)	+3.70		
(cyclohexane ring with F, H)	+49.0	^2J	−12.6	(cyclohexane ring with H, H)	

[a]Reproduced with permission from Wiley Heydon Ltd.[2]

consideration of the magnetogyric ratios[6] of ^{13}C and ^{1}H (which are in a ratio of 1:4 and accordingly predict J_{CH}/J_{HH} to be 0.25) is inadequate to describe the relative magnitudes of the couplings of two different types of nuclei.[8] Notwithstanding the lack of understanding of why the ratio J_{CH}/J_{HH} is larger than anticipated, the good correspondence between J_{CH} and J_{HH} argues for similar coupling mechanisms in these two types of couplings. A ready interpretation for this behavior is that the same spin number, the similar size, and the similar electron configurations (no nonbonding electrons) of ^{13}C and ^{1}H are sufficient to ensure similar mechanistic contributions.

A Check for the Reliability of the J_{CH}/J_{HH} Correlation: Dihydroaromatics

Because the correlation of Figure 2-1 extends reliably through a nearly comprehensive series of structural types, it might be expected that J_{CH} values predicted from proton-proton models could serve as a dependable tool for conformational and structural analysis in new systems not previously tested. An opportunity to check the usefulness of this expectation was afforded in the conformational analysis of dihydroaromatics, where the homoallylic $^5J_{HH}$ value had been used in an attempt to establish the conformation of these compounds. Specifically, it was not known which of the three types of dihydroaromatics 20–22 were puckered and which were flat.

Before the carbon–proton coupling study was conducted, the situation was as follows. Most evidence indicated that dihydrobenzenes (20) were flat,[9] although some disagreement was voiced.[10] The conformation of 22 was certainly puckered with the substituent axial.[11] Scanty data suggested that dihydronaphthalenes (21) were puckered with the substitutent axial,[12] but the degree of puckering was uncertain. The difficulty in the conformational analysis of compounds 20–22 lay in the uncertainty of the usefulness of the homoallylic proton–proton coupling constants (for the protons in question, see compounds 20–22). The cis–trans coupling ratios were about the same for each of the compounds **20–22**. It was uncertain whether this similarity was a result of (1) the unreliability of homoallylic couplings in conformational analysis in general, or (2) the expectation that equatorial–equatorial couplings should be similar to equatorial–axial couplings. There was valid reason for this second expectation because the equatorial protons cannot couple efficiently across the ring (the hyperconjugative mechanism operating through the π cloud has maximum overlap with an axial substituent and minimum overlap with an equatorial substituent). Compounds **23** and **24** demonstrate these principles. In compound **23**, which is flat, of course cis and trans couplings should be similar, because

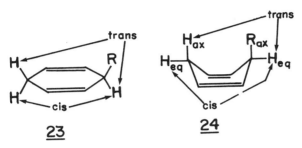

all protons can interact with the π cloud with the same efficiency. In compound **24**, which is puckered, the cis and trans couplings each involve an equatorial proton and accordingly both these couplings should be small and may be approximately the same. Thus, the cis–trans homoallylic coupling constant ratio may be close to unity irrespective of the geometry of the dihydroaromatic ring.

To resolve this dilemma, carbon–proton couplings were utilized.[13] It was recognized that axial–axial and axial–equatorial couplings were better suited to the conformational analysis of **20–22**. Because axial couplings are expected to be larger than equatorial couplings, then J_{ax-ax} should be substantially larger than J_{ax-eq}, and the J_{CH}/J_{CH} ratio need not remain so close to 1. Because the substituent is axial, then couplings to the substituent must be used; accordingly, the substituent was labeled with ^{13}C. In this study, the compounds **25–27** were synthesized and deuterium-decoupled proton NMR spectra were analyzed (the deuteriums were placed in the molecules to simplify the analysis and to avoid the deceptively simple spectral problems of dihydrobenzenes). The results are shown in Table 2-3.

For dihydrobenzoic acid (**25**) the two homoallylic J_{CH} values are about the same, and for **26** and **27** they progressively become much different. Specifically, the $J_{CH}(cis)/J_{CH}(trans)$ ratio steadily progresses from 1.24 for **25** to 4.6 for **27**, clearly indicating the carbon substituent is becoming more nearly axial. Further, the $J_{CH}(cis)/J_{CH}(trans) = 1.24$ in **25** is virtually the same as the corresponding proton–proton ratio $J_{HH}(cis)/J_{HH}(trans) = 1.22$, clearly supporting the predominating view that monosubstituted dihydrobenzenes (**20**) are flat. The conclusion from the data of Table 2-3 is manifest: Both proton–proton and

TABLE 2-3. CARBON–PROTON AND PROTON–PROTON COUPLING CONSTANTS (Hz) FOR DIHYDROAROMATIC CARBOXYLIC ACIDS **25–27**[a]

Compound	cis-$^5J_{HH}$	trans-$^5J_{HH}$	cis-$^5J_{CH}$	trans-$^5J_{CH}$	$^2J_{HH}$	$^2J_{CH}$
25	9.19	7.56	5.75	4.65	(−22.0)	(−)10.97
26	3.84	4.36	5.44	2.86	(−21.9)	(−)9.85
27	<0.5	0.9	3.2	0.7	−18.1	(−)9.0

[a]The chemical shifts of proton (1) are 3.705, 4.30, and 4.99, respectively. The chemical shifts of proton (4c) are 2.574, 3.34, and 4.25, respectively. The chemical shifts of proton (4t) are 2.559, 3.19, and 3.90, respectively. (Reprinted with permission from ref 13. Copyright 1982 American Chemical Society).

carbon–proton couplings in the dihydroaromatic series **25–27** show a monotonic increase in the amount of puckering, with the carbon–proton data being the more reliable because J_{CH} involves an axial–axial coupling.

The final vindication of the method of carbon–proton couplings as a conformational tool would be the appropriate values of Table 2-3 fitting on the plot of Figure 2-1, i.e., the J_{CH}/J_{HH} ratio for geometrically equivalent systems being about 0.62. Indeed, these ratios for **25** are calculated to be: $J_{CH}(cis)/J_{HH}(cis) = 5.75/9.19 = 0.63$, and $J_{CH}(trans)/J_{HH}(trans) = 4.65/7.56 = 0.62$, in excellent agreement with the slope of Figure 2-1! Thus, a nice, self-consistent picture emerges that suggests that proton–proton couplings can be used reliably as models for carbon–proton couplings in homoallylic geometries, systems which were not included in Figure 2-1.

Aliphatic Couplings

Early Work

The thought seemed natural to several investigators that C—H couplings might be analogous to proton–proton couplings, and for the past 20 years numerous efforts have been applied to explore this possibility. The early work of Karabatsos[14-19] involved the determination of various aliphatic $^3J_{CH}$ and $^2J_{CH}$ values in simpler aliphatic derivatives. He was the first to note that in general $^3J_{CH}$ is greater than $^2J_{CH}$,[14,16] although the "reasons were not clear why." He attempted a plot of $J_{C-C-C-H}$ vs. $J_{H-C-C-H}$ for a number of isopropyl and tert-butyl compounds and observed[14] for alcohols, ketones, and hydrocarbons (seven points) a correlation appeared to exist although the "meaning and significance was debatable." He saw that both two-bonded and three-bonded C—H couplings were less than the analogous H—H couplings. He further recognized that the hybridization of the terminal carbon was important; in general, the greater the s character the greater the three-bonded coupling in a series of tert-butyl compounds of the general structure $(CH_3)_3CCH_2X$, $(CH_3)_3COR$, $(CH_3)_3CCN$:[17,18] 3.6–6.0 for sp^3, 3.7–6.4 for sp^2, and 5.4 for sp hybridization. However, he recognized an "inadequacy" of this correlation in that there were unusually high $^3J_{CH}$ values for halogens, and the increase in these values was in the order of $Cl < Br < I$, whereas from halogen electronegativities the reverse order was expected. From these "anomalies," it was hypothesized that spin-dipole and/or electron-orbital contributions might be occurring[17] in addition to the Fermi contact mechanism. It was implied that these variances with proton–proton couplings might portend difficulties in attempting to correlate C—H couplings with H—H couplings; however, subsequent work in the past 15 years has not required the inclusion of mechanistic contributions other than the Fermi contact term.

Undeniably the greatest use of proton–proton couplings in conformational analysis has been in the development and application of the Karplus relationship[5] (dihedral angular dependence), wherein $^3J_{HH}$ exhibits two maxima when the three-bonded dihedral angle of the H—C—C—H framework is 0° and 180° (the smaller maximum is at 0°) and a minimum at 90°. Accordingly, $^3J_{HH}$ can vary perhaps 10 Hz (at 0°) to 0 Hz (at 90°) to 14 Hz (at 180°). It was obvious to attempt such a relationship for C—H couplings. Early on, Karabatsos saw[19] such a difference of $^3J_{CH}$ in the gauche and trans conformations of propionaldehyde and its oxime O-methyl ether. The gauche (28) and trans (29) populations of propionaldehyde and the gauche (30) and trans (31) populations

28
gauche

$J_t(HH) = 7.2$

$J_g(CH) = 0.2$

29
trans

$J_g(HH) = 0.2$

$J_t(CH) = 3.5$

30
gauche

$J_t(HH) = 10.9$

$J_g(CH) = 0.7$

31
trans

$J_g(HH) = 2.8$

$J_t(CH) = 7.8$

of propionaldehyde oxime O-methyl ether were already known as a function of temperature, and from plots of the observed $^3J_{HH}$ and $^3J_{CH}$ values of these compounds vs. temperature the respective proton–proton and carbon–proton couplings were calculated (see **28–31**). The respective comparisons were: for propionaldehyde, $J_g(CH)/J_g(HH) = 0.2/0.2$ and $J_t(CH)/J_t(HH) = 3.5/7.2$; for propionaldehyde oxime O-methyl ether, $J_g(CH)/J_g(HH) = 0.7/2.8$ and $J_t(CH)/J_t(HH) = 7.8/10.9$. Therefore, it was apparent that in these compounds the C—H couplings, as expected, were less than the geometrically analogous H—H couplings but exhibited similar dihedral angular dependence.

General Aspects of Aliphatic $^{13}C—^1H$ *Couplings*

The propane moiety **32** where X is various electronegative and electropositive substituents has been studied in detail, and the results suggest the existence of

$$CH_3—CH—CH_3$$
$$|$$
$$X$$

32

general trends in C—H couplings.[20] For $^2J_{C-2,H}$ (see **33**), the values do not vary significantly (-4.0 to -4.6 Hz); however, for $^2J_{C-1,H}$ (see **34**), the values vary

$$^2J_{C-2,H}$$ $$^2J_{C-1,H}$$

33   **34**

considerably (from -10.31 to -5.0 Hz). All these geminal couplings are assumed to be negative, as in propane.[21] These trends have been verified in other systems,[22] and negative $^2J_{CH}$ values have been reported in ethyl derivatives.[23-26] Therefore, $^2J_{CH}$ values are relatively unaffected when the coupling carbon is substituted, but they change dramatically when the substituent is situated on the carbon that intervenes between the coupling nuclei, i.e., C—CX—H. Proton–proton couplings likewise vary considerably as the carbon between the coupling nuclei is substituted, i.e., H—CX—H.[27] Both $^2J_{CH}$ and $^2J_{HH}$[27] are generally negative.

For $^3J_{CH}$ couplings in propane **32**, the values decrease somewhat as the electronegativity of the substituent increases.[20] This is to be compared with terminally substituted compounds (such as $CH_3CH_2CH_2OH$) where an electronegative substituent causes an increase in $^3J_{CH}$.[16] Therefore, an internal electronegative substituent (see **35**) decreases $^3J_{CH}$ and a terminally substituted

$$\underset{\textbf{35}}{\overset{\overset{\displaystyle H}{|}}{C}{-}C{-}\underset{\underset{\displaystyle X}{|}}{C}} \qquad \underset{\textbf{36}}{\overset{\overset{\displaystyle H}{|}}{C}{-}\underset{\underset{\displaystyle X}{|}}{C}{-}C} \qquad \underset{\textbf{37}}{\overset{\overset{\displaystyle H}{|}}{C}{-}\underset{\underset{\displaystyle X}{|}}{C}{-}H}$$

substituent (see **36**) increases $^3J_{CH}$, and an alternating substituent effect prevails in accord with semiempirical molecular orbital (MO) calculations.[28] The internal substituent effect in **35** parallels the geometrically equivalent proton–proton system **37**, where an electronegative substituent likewise decreases $^3J_{HH}$. Both $^3J_{CH}$[21] and $^3H_{HH}$[27] values are generally positive.

To test the idea that aliphatic $^3J_{CH}$ values parallel geometrically equivalent $^3J_{HH}$ values, a study of systems **38** and **39** was conducted.[29] In both the *tert*-butyl system **38**, which affords the $^3J_{CH}$ values, and the isopropyl system **39**, which gives the $^3J_{HH}$ values, each coupling arises from the average of two gauche and one anti conformation, and geometrical differences do not exist between the C—H and the H—H systems. A plot of $^3J_{CH}$ vs. $^3J_{HH}$ for the respective substituents afforded a straight line with a correlation coefficient of 0.991, from

$$\underset{\textbf{38}}{\overset{\overset{\displaystyle H \qquad CH_3}{|\qquad\quad|}}{CH_2}{-}\underset{\underset{\displaystyle CH_3}{|}}{C}{-}X} \qquad\qquad \underset{\textbf{39}}{\overset{\overset{\displaystyle H \qquad CH_3}{|\qquad\quad|}}{CH_2}{-}\underset{\underset{\displaystyle H}{|}}{C}{-}X}$$

which the relationship $^3J_{CH} = 1.2$ Hz $\times$ $^3J_{HH} - 3.5$ Hz could be derived. This correlation indicated to the authors the existence of "parallel behavior of carbon–proton and proton–proton couplings toward substituent electronegativity." Specifically, both $^3J_{CH}$ and $^3J_{HH}$ values for the internally substituted systems **38** and **39** decrease as the electronegativity increases in the series X = D, CH_3, I, Br, Cl, and F.

These findings clearly indicate C—H and H—H couplings behave similarly and attest to the similar operating coupling mechanisms in the two cases. However, one should be careful when utilizing J_{CH}/J_{HH} correlations to be certain that the same geometries exist for the corresponding C—H and H—H systems under consideration. For example, in the original correlation of $^3J_{CH}$ vs. $^3J_{HH}$ in isopropyl compounds[14] a correlation existed that was inferior to the *tert*-butyl/isopropyl study cited above, possibly because the carbon–proton and proton–proton couplings were taken from the same compounds. A problem might exist owing to the differences in the orientation of the substituent with respect to the coupling nuclei (see **40**). The Newman projections **41a** and **41b** indicate that the orientation of the carbonyl is different with respect to the proton in **41a** (giving $^3J_{HH}$) from its orientation with respect to the methyl group in **41b** (giving $^3J_{CH}$). If orientation of the substituent is important to couplings

40   **41a**   **41b**

(see $^3J_{CC}$ vs. Orientation of a Terminal Substituent, Chapter 3), then it can be seen that the proton–proton model in **41a** and the carbon–proton model in **41b** cannot be expected to be equivalent. Even more serious is the case where the geometries of the C—C—C—H and the H—C—C—H linkages themselves are different. For example, one hardly expects the $^3J_{CH}$ couplings in **42** ($J_{CH_3-H_{ax}}$ and $J_{CH_3-H_{eq}}$) to correlate with the respective $^3J_{HH}$ couplings ($J_{H_1-H_{ax}}$ and $J_{H_1-H_{eq}}$), because the former involves equatorial–equatorial and equatorial–axial couplings and the latter involves axial–axial and axial–equatorial couplings.

42

The Karplus Dihedral Angle Relationship

Without a doubt the Karplus dihedral angle relationship[5] has been the strongest tool in conformational analysis by proton NMR, and it is natural to attempt an analogous correlation in carbon–proton couplings. Numerous examples attest to similar behavior for carbon–proton couplings, and Karplus-like dihedral angle plots have been constructed from both theoretical and empirical data.

In a theoretical study for propane,[30] the Fermi contact contribution to three-bonded C—H coupling was calculated. The resulting data (Table 2-4) predict maxima of $^3J_{CH}$ at 0° and 180° of 6.82 and 8.81 Hz, respectively, and a minimum near 90° of 0.68 Hz, from which the following equation was derived:

$$^3J_{CH} = 4.26 - 1.00 \cos\theta + 3.56 \cos 2\theta$$

A number of applications of the dihedral angular dependence of vicinal carbon–proton couplings has been made, notably in biochemical system. In

TABLE 2-4. CALCULATED $^3J_{CH}$ VALUES FOR
PROPANE VS. DIHEDRAL ANGLE OF THE INVOLVED
C—C—C—H FRAMEWORK[a]

Angle (degrees)	$^3J_{CH}$ (Hz)
0	6.82
15	6.36
30	5.15
35	4.63
45	3.54
60	2.00
75	0.94
85	0.68
90	0.70
105	1.41
120	2.95
135	4.96
150	6.91
155	7.46
165	8.31
180	8.81

[a]Reproduced with permission.[28]

uridine (43) derivatives, eight values of $^3J_{C-H}$ ranging from 2.0 to 8.7 Hz were observed for estimated dihedral angles varying from 50° to 180°.[31] In another study involving these uridine derivatives, more accurate dihedral angle values were obtained from X-ray analysis, and the existence of a Karplus-like relationship was verified.[32] The authors of this latter study state that two values "may be denoted as anomalies to this Karplus curve" (see Figure 2-2) and that "care should be exercised in the interpretation of NMR results in cases where anomalous bonding is concerned," but in view of substituent and through-space effects (see Chapter 3) the relationship shown in Figure 2-2 can be considered quite good and should attest to the reliability of the Karplus-like relationship in carbon–proton couplings. Indeed, "irregularities" of the sort shown in Figure 2-2 should not be considered anomalous but normal (for that matter, perhaps one

43

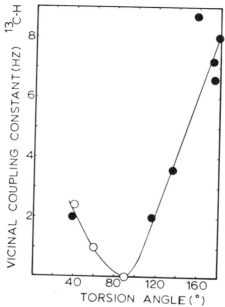

Figure 2-2. Plot of the coupling constants vs. torsion angles in uridine derivatives. The open circle at 40° is for uridine, at 60° for 2-phenyl-5-methyl-1,3-dioxane, and at 90° for α-D-glucopyranose triacetate 1,2-(methyl orthoacetate). (Reproduced with permission.[32])

of the other values off the curve near 180° is normal and the other three are "irregular").

The Karplus-like relationship has also been applied to carbohydrates. An octadeuterio derivative of methyl β-cellobioside (**44**) was synthesized and studied by ^{13}C NMR.[33] X-Ray crystallographic data were previously obtained for **44**, and accordingly accurate estimates of dihedral angles were available. The generated $^3J_{CH}$ values were plotted as shown in Figure 2-3, and the points again reflect the general Karplus plot. Apparently the Karplus-like relationship can be applied to a wide variety of organic systems.

The Karplus-like relationship also has been used in carbon–proton couplings in flexible systems where the carbon has been functionalized as a cyanide

$$\underline{44}$$

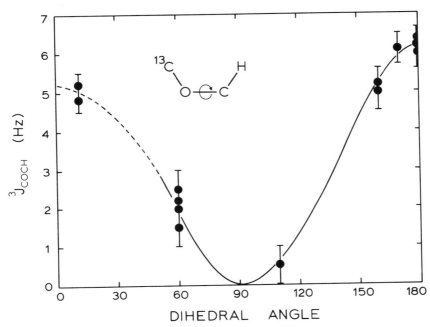

Figure 2-3. Plot of the coupling constants vs. torsion angles in β-methylcellobioside. (Reproduced with permission.[33])

group.[34] In **45** the COOH—H_3 carbon–proton coupling is 2 Hz, indicating a gauche geometry, whereas the CN—H_3 carbon–proton coupling of 6 Hz indicates an "average" coupling. Therefore, the CN group is gauche to H—3 and trans to H—5, but in the mirror image the CN group is trans to H—3 and gauche to H—5. The true trans value of $^3J_{CN,H}$ should be 9 Hz, as taken from the more rigid cyclohexanone compound **46**, and the observed 6 Hz coupling in **45** is roughly an average of 9 (trans) and 2 (gauche) Hz.

The values of $^3J_{CH}$ in the literature vs. their respective dihedral angles are remarkably in agreement with the theoretical values of Table 2-4, considering substituent and hybridation variations. In the uridine studies cited above,[31, 32] the $^3J_{C-H}$ values at 180° were about 8 Hz, to be compared with the calculated value of 8.8 Hz. In the methyl β-cellubioside case[33] the 180° and 0° J_{C-H} values were ca. 6 Hz and 5 Hz, respectively, comparing with the respective theoretical values of 8.8 Hz and 6.8 Hz. For the CN—H couplings[34] the 180° and 60° J_{C-H} values were 9 and 2 Hz, respectively, to be compared with the theoretical values of 8.8 and 2.0 Hz, respectively. In other studies of glucose derivatives and other monosaccharide compounds[35] $^3J_{C-H}$ values of 0–3 Hz were noted for a dihedral angle of 60–100° and of 4.5–5.5 Hz for 140–180°, to compare with the theoretical values of 0–2.0 Hz and 5–8.8 Hz, respectively. For glucopyranosides[36] this established relationship has been used to help evaluate dihedral angles. This relationship also has been utilized in the conformational analysis of 2,6-dichloro-1,4-oxathianes, where J_{CH} values were used to determine the conformations of both isomers,[37] and also in the assignment of the glycosylation site on nitrogen heterocycles.[38] For these last two cases, the conformations were confirmed using geminal J_{CH} values (see discussion in Geminal J Values in Conformational Analysis, this chapter). It can be concluded that the Karplus-like relationship is well established for carbon–proton couplings. Proton–proton couplings, with the myriad of past successes in conformational analysis, were never any more accurate than carbon–proton couplings in the determination of molecular geometry.

Conformational Analysis of Amino Acids

The conformational analysis of amino acids well exemplifies the power and utility of carbon–proton couplings in the determination of molecular geometry. The study of amino acids of the general structure **47** involves the determination of the three conformers **47t**, **47g**, and **47h** (for trans, gauche, and hindered, respectively).[39] The successful proton NMR analysis of **47** rests on the correct

$$\text{COOH}$$
$$|$$
$$\text{H}_2\text{N}-\text{C}-\text{H}$$
$$|$$
$$\text{CH}_2\text{R}$$

47

47t   47g   47h

interpretation of the ABC pattern (of H_A, H_B, H_C) in the proton NMR. The difficulty resides in the ambiguity of proton assignments of H_A and H_B, whose chemical shifts can be quite similar. The total amount of the three conformers is

$$(t + g + h) = 1 \tag{1}$$

and the observed coupling constants can be related to the true general coupling constants J_G (gauche coupling) and J_T (trans coupling) as follows:

$$J_{AC} = tJ_G + gJ_T + hJ_G \tag{2}$$

$$J_{BC} = tJ_T + gJ_G + hJ_G \tag{3}$$

If the general gauche and trans couplings J_G and J_T are known, and because J_{AC} and J_{BC} can be measured, the solution of eqs. (1)–(3) will give the fractional populations of t, g, and h.

In general analysis of the equations above, eq. (2) can be subtracted from eq. (3) affording:

$$J_{BC} - J_{AC} = (t - g)(J_T - J_G) \tag{4}$$

Certainly $J_T > J_G$; and if $J_{BC} > J_{AC}$, then $t > g$. However, if the assignments of H_A and H_B are reversed (refer to the ambiguity above), then the calculated population of t and g are exchanged. One would expect $t > g$, because less steric hindrance is expected in t, but the uncertainty remains. Thus, one is faced with the certain determination of h and the tentative determination of t and g.

The ambiguity is removed by utilizing carbon–proton couplings. Equations similar to (2) and (3) can be developed involving the carbon atom X^{39} (K symbolizes J_{CH}):

$$K_{XA} = tK_G + gK_G + hK_T \tag{5}$$

$$K_{XB} = tK_G + gK_T + hK_G \tag{6}$$

and

$$K_{XB} - K_{XA} = (g - h)(K_T - K_G) \tag{7}$$

As proton–proton couplings gave only h unambiguously, carbon–proton couplings will give t unambiguously. Therefore, from eq. (1) all three values will be known.

Values for J_T, J_G, K_T, and K_G have been determined for a number of compounds of the general structure **47** (phenylsuccinic acid, histidine, cysteine, aspartic acid, alanine, valine, homoserine, and derivatives thereof) and their ranges are:[39-42]

$$J_T = 12.8–13.5$$

$$J_G = 2.5–2.8$$

$$K_T = 9.8–11.9$$

$$K_G = 0.4–2 \tag{8}$$

The ranges of each of the J and K values probably stem from uncertainties in the determinations and from inherent differences among the systems studied. The values in eq. (8) and eqs. (4) and (7) allow unequivocal determination of t, g, and h even when H_A and H_B cannot be unequivocally assigned. For example the relative populations for aspartic acid (**47**, $R = CO_2H$) are $t = 0.62$, $g = 0.15$, $h = 0.23$.[40]

In this conformational analysis of amino acids, two complications may arise. First, some amino acids do not have two β protons and accordingly have the general structure **48**, such as valine, where $R = CH_3$. In this case the system is

$$
\begin{array}{c}
CO_2H \\
| \\
H_2N-C-H \\
| \\
R-C-H \\
| \\
R
\end{array}
$$

48

ABX (for H_A, H_B, C_X). The observed J_{HH} value allows only the determination of the fractional population of **48a** because:[42]

$$[\mathbf{48a}] = (J_{AB} - J_G)/(J_T - J_G) \tag{9}$$

However, the observed K_{CH} value allows the additional determination of the fractional population of **48c**:[42]

$$[\mathbf{48c}] = (K_{XA} - K_G)/(K_T - K_G) \tag{10}$$

The fractional populations of **48a**, **48b**, and **48c** are thereby determined.

48a   **48b**   **48c**

The second complication in the general analysis of amino acids occurs when for the general structure **47** the chemical shifts of A and B are identical, and the resulting proton system is now classified as AA′C. Most amino acids at low pH have this property.[42] Deceptively simple proton spectra are observed for these systems, where only the sum of J_{AC} and $J_{A'C}$ can be obtained (the AA′ portion appears as a triplet), and only the value **47h** can be determined from eq. 11, derived by adding eqs. (2) and (3), substitution of eq. (1), and rearranging:[42]

$$\mathbf{47h} = \frac{(J_G + J_T) - (J_{AC} + J_{A'C})}{J_T - J_G} \tag{11}$$

However, the observed K_{AX} can give the value of **47t** [from eqs. (5), (6), and (1)]:[42]

$$47t = \frac{(K_G + J_T) - (J_{AX} + J_{A'X})}{K_T - K_G} \tag{12}$$

Thus are obtained t, g, and h from eq. (1).

Geminal J Values in Conformational Analysis

Applications of $^2J_{HH}$ values have been limited principally to the determination of the angle between the geminal protons (H—C—H) and to the determination of the orientation of the H—C—H moiety to an adjacent π system or to α substituents.[27] Carbon–proton couplings appear to follow the trends so far observed in proton–proton couplings. First, the smaller the angle of the carbon bonds of the methylene group (and thereby presumably the larger the H—C—H angle), the larger (less negative) the J_{CH} value (Table 2-1), precisely as is observed for J_{HH} values. Second, $^2J_{CH}$ couplings in dihydroaromatics (Table 2-3) become larger (less negative) as the ring becomes more puckered, exactly as is observed[27] and predicted[43] for proton–proton couplings (the $^2J_{HH}$ value is predicted to be most negative when the hyperconjugative effect is maximum, i.e., when the C—H bonds most nearly parallel the p orbitals as in **49**[43]).

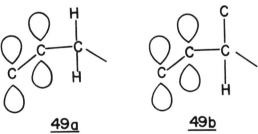

49a   **49b**

A strikingly new use of geminal C—H couplings in conformational analysis has involved carbohydrates.[35,44–47] Initial C—H coupling studies in 1969 suggested a method for differentiating between α and β isomers of D-glucose.[44] Specifically ^{13}C-enriched monosaccharides (glucose, idose, and related lactones) in a ^{1}H NMR study[35] and natural abundance monosaccharides and disaccharides (psicopyranose, β-cellobiose, maltose, mannofuranose, and derivatives thereof) in a ^{13}C NMR study[45] clearly showed the orientation of an oxygen substituent influences greatly a geminal C—H coupling. More recent studies again involve ^{1}H NMR for ^{13}C-labeled compounds utilizing double-irradiation[46] and ^{13}C NMR for natural abundance compounds using the "selective population transfer" technique.[47] This research has enabled the *signs* of the C—H couplings to be determined, and the results are quite interesting. In a very general sense, when an oxygen is antiperiplanar to a C—C—H linkage (see **50**), the geminal C—H coupling is large and positive; when the oxygen is of opposite orientation (**51**), the geminal C—H coupling is negative (but with a

$^2J_{CH}$ is positive $^2J_{CH}$ is negative

50   **51**

comparable absolute value).[46] Specifically, for β-D-glucopyranose-1-^{13}C (**52**) the $^2J_{C-H_2}$ value is -5.7 Hz, for α-D-glucopyranose-1-^{13}C (**53**) the $^2J_{C-H_2}$ value is small (<1 Hz), and for 1,2:5,6-di-*O*-isopropylidene-α-D-glucofuranose-1-^{13}C (**54**) the $^2J_{C-H_2}$ value is $+5.5$ Hz. The Newman projections pertinent to the $^2J_{CH}$

couplings in **52–54** are **55–57**, respectively. In **55** the "vector resultant" (v.r.) of O_5 and O_1 eclipses the C_1-H_2 bond ($\theta=0°$) and the model **51** would predict a negative $^2J_{CH}$ ($^2J_{obs} = -5.7$ Hz). In **57** the vector resultant of O_5 and O_1 is anti

55   **56**   **57**

$\theta = 0°$ $\theta = 180°$

to the C_1-H_2 bond ($\theta=180°$) and the model **50** predicts a positive coupling ($^2J_{obs} = +5.5$ Hz). In **56** the resultant vector of O_5 and O_1 is intermediate ($\theta = 120°$) and the observed $^2J_{CH}$ is approximately midway between the two extremes ($^2J_{C-H_2} < 1$ Hz). This general trend is quite reminiscent of the $^2J_{CH}$ couplings in substituted olefins—in **58** the anti coupling $^2J_{C_1-H_A}$ is $+7.5$ Hz and the eclipsed coupling $^2J_{C_1-H_B}$ is -8.5 Hz.[48,49]

58

The observations discussed above also appear to apply when only one oxygen is substituted to the terminal carbon, as portrayed in **50** and **51**, irrespective of whether one or two oxygens is substituted internally, as portrayed in **59** and **60**.

59   **60**

A large number of compounds have been studied and the data continue to bear out the general phenomenon[47]: In the linkage O—C—C—H, when the oxygen

TABLE 2-5. CALCULATED $^2J_{CH}$ VALUES VS. THE DIHEDRAL ANGLE OF O—C—C—H (OR RESULTANT VECTOR) IN ETHANOL AND 1,1-ETHANEDIOL[a]

RESULTANT VECTOR OF TWO –OH GROUPS

Ethanol J_{CH} (Hz)	Angle (degrees)	1,1-Ethanediol J_{CH} (Hz)
−3.2	0	−3.0
−3.0	30	−2.8
−2.2	60	−1.9
0	90	0
+3.9	120	+2.6
+7.2	150	+4.9
+9.3	180	+6.0

[a]Reproduced with permission.[46]

is anti to the proton, the contribution to $^2J_{CH}$ is positive ($^2J_{CH}$ can vary between 0 and +8 Hz), and when the oxygen is gauche to the proton, the contribution to $^2J_{CH}$ is negative ($^2J_{CH}$ can vary between 0 and −5 Hz).

The results of recent calculations reinforce the trends shown by the general empirical data nicely. The calculations for ethanol and 1,1-ethanediol are shown in Table 2-5.[46] Furthermore, calculations for derivatives substituted with more oxygen substituents reflect empirical data.[47] Specifically, Table 2-6 shows the ranges for three different polyols, which quite closely parallel observed data (viz., compounds of the general structure **62** give data varying from ca. 0 to +8 Hz, and compounds of the general structure **63** give data with a range of ca. 0 to −5 Hz).

In summary, geminal carbon–proton couplings appear to portray accurately the orientation of the oxygen substituent in the linkage O—C—C—H. A similar trend may exist for other electronegative substituents, such as chlorine.[47] Conformational analysis using geminal carbon–proton couplings related to the

TABLE 2-6. Calculated Ranges of $^2J_{CH}$ vs. the Dihedral Angle of the Resultant Vector in Various Polyols[a]

Compound	$^2J_{CH}$	
	$\theta = 0°$	$\theta = 180°$
HO⧵ ⧸H H—^{13}C—C—OH HO⧸ ⧵CH$_3$ **61**	−6	+4
CH$_3$⧵ ⧸H H—^{13}C—C—OH OH⧸ ⧵OH **62**	−2	+8
CH$_3$⧵ ⧸H H—^{13}C—C—OH HO⧸ ⧵CH$_3$ **63**	−9	0

[a]Reproduced with permission.[47]

orientation of an electronegative substituent promises to be a powerful tool in the elucidation of molecular structure.

Olefinic Couplings

A complete analysis of propene (**64**)[50] points the way for general trends in the magnitudes of C—H couplings in olefins. First, the terminal geminal couplings (**64a**) are modest and positive (4.95 Hz), just as for geminal proton–proton couplings.[27] Second, the vicinal C—H couplings across the olefin (**64b**) are larger and positive, and the trans couplings are larger than the cis couplings (+12.65 and +7.55 Hz, respectively), just as for geometrically equivalent H—H couplings.[27] Third, the internal geminal couplings through the olefin (**64c**) can be quite variable, either negative or positive; of course there is no equivalent proton–proton model because a proton cannot be doubly bonded. Fourth, internal geminal couplings without the olefin (**64d**) are negative (−6.75 Hz); these couplings are akin to aliphatic geminal couplings. Finally, the terminal vicinal couplings (**64e**) are substantial and positive (+6.70); these again appear similar to aliphatic vicinal couplings.

Vicinal Couplings

There are extensive data in the literature describing cis–trans C—H couplings.[51-53] As for H—H couplings, an internally substituted electronegative group decreases both the cis and the trans vicinal couplings. Hence, for

2-bromopropene (**65**) these couplings are substantially reduced (compare with **64b**).[52] For externally substituted olefins (see 3-bromopropene, **66**), the cis–trans couplings are increased,[52] as was observed for terminally substituted aliphatic compounds.

Hybridization effects for cis–trans C—H couplings are similar to such effects in aliphatic couplings. For example, for compounds **67–69** both cis and trans couplings increase as the s character increases on the carbon.[52]

Bond order also plays an important role in vicinal olefinic C—H couplings. In **70** it is seen that vicinal couplings are larger for coupling paths that traverse the entire olefin.[52]

A comparison of carboxylic acids (see **71**) and aldehydes (**72**)[52] at first sight suggests a breakdown of the rule that terminal electronegative substituents cause an increase of $^3J_{CH}$: in **71** the couplings are smaller than in **72** notwithstanding an additional OH substituent. However, as studies on carbon–carbon couplings have shown (see $^3J_{CC}$ vs. Orientation of a Terminal Substituent, Chapter 3) the carbonyl group exerts a considerable negative through-space contribution to the coupling. In the aldehyde, the time-averaged conformation of the substituent is different because the carboxaldehyde proton in **72** is smaller than the hydroxyl group in **71** and this proton can point toward the cis proton, thereby preventing carbonyl through-space effects.

71   **72**

Steric effects can play a significant role in determining the value of $^3J_{CH}$, particularly trans C—H three-bonded couplings.[52] Compounds **73–75** illustrate

73   **74**   **75**

this effect: incorporation of the β-methyl group in **74** and **75** notably decreases trans-$^3J_{CH_3,H}$ in **73** from 10.1 to 7.7 Hz. In contrast, the trans-$^3J_{COOH,H}$ coupling in **73** (12.8 Hz) changes little with the incorporation of the β-methyl group in **75** (13.2 Hz). Apparently this steric effect occurs whenever the carbon involved in the trans C—H coupling bears at least one hydrogen. For example, the trans $^3J_{CHO,H}$ coupling of 15.2 Hz in **76** drops to 11.0 Hz in **77**, and the trans $^3J_{CH_2OH,H}$

76   **77**

coupling of 10.8 Hz in **78** decreases to 9.3 Hz in **79**. When the coupling carbon is —COOH, —COOR, or —CN, no such steric effect is observed.[52] Once the

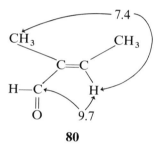

78   **79**

steric effect is recognized, then tigaldehyde (**80**) is no longer regarded as an anomaly—"some uncertainty exists in the assignment of splittings ..."—but instead as a case where the cis-$^3J_{CHO,H}$ value of 9.7 Hz is normal and the trans-$^3J_{CH_3,H}$ value of 7.4 Hz is perfectly in accord with that in **74**.

80

Table 2-7 lists representative $^3J_{CH}$ values for simpler molecules. Isopropenyl compounds are included because of the wide variety of functionality and because steric effects should be minimal. The $^3J_{CH}$ values in this table follow trends discussed above.

TABLE 2-7. VICINAL CARBON–PROTON COUPLINGS IN SIMPLE OLEFINS

Compound	cis-$^3J_{CH}$	trans-$^3J_{CH}$	Reference
$H_2C{=}CH{-}CH_3$	7.6	12.6	50
$H_2C{=}CH{-}CH_2Br$	9.1	15.5	52
$H_2C{=}CH{-}CHO$	10.1	15.9	52
$H_2C{=}CH{-}COOH$	7.6	14.1	52
$H_2C{=}CClCH_3$	4.6	8.9	52
$H_2C{=}CBrCH_3$	4.1	8.7	52
$H_2C{=}C(CH_3)CH(CH_3)_2$	6.9	11.3	52
$H_2C{=}C(CH_3)CH_2OH$	6.2	10.8	52
$H_2C{=}C(CH_3)CH_2Cl$	7.6	13.8	52
$H_2C{=}C(CH_3)CHO$	9.4	15.2	52
$H_2C{=}C(CH_3)COOH$	6.5	12.8	52
$H_2C{=}C(CH_3)C{\equiv}CH$	8.1	14.7	52
$H_2C{=}C(CH_3)CN$	8.5	15.7	52

Assignment of Structure Using $^3J_{CH}$ Values

The general trends noted for olefinic carbon–proton couplings have been used for structure assignments. In the case of α_2-guttiferin (81) the $^3J_{CH_3,H}$ value of 7.3 Hz indicates a cis coupling and the $^3J_{COOH,H}$ value of 13.1 Hz indicates a

81

trans coupling (compare with 64b and 71).[52] If one is concerned lest steric effects could invalidate this approach, 73 can be compared with structurally similar tiglic acid (74) and angelic acid (75). The values of 81 closely resemble those of 75 and clearly establish the structure of 81 as Z, just as in angelic acid (75).

82

The geometry of phenylitaconic acid (82) was also established using C—H couplings.[53] The methylene carbon coupled to the olefinic proton with $J_{CH} = 8.1$ Hz, clearly a trans coupling (with the steric effect), whereas the carboxyl group coupled to the olefinic proton with $J_{CH} = 6.7$ Hz, obviously a cis coupling. These relationships of the olefinic proton with the methylene and carboxyl group establish the geometry of 82 as E.

The power of C—H couplings can be illustrated in the analysis of mixtures. Upon condensation of methyl acetoacetate with benzaldehyde, an inseparable mixture of 83 and 84 results. Proton NMR would be worthless for quantitative analysis unless secure assignments could be made for the signals, a precarious task because proton–proton couplings would be of no help (all aliphatic signals are singlets) and because chemical shift assignments of the isolated nuclei would be difficult. Carbon–proton couplings, however, render analysis easy: Respec-

83   **84**

tive cis and trans couplings are readily identifiable, and the major product could be recognized as **83**.[53]

It is true that carbon–proton couplings involve greater complexity than proton–proton couplings, but within this manifold behavior of C—H couplings lies its strength. Indeed, the behavior of C—H couplings, once understood, is no more random than is that of H—H couplings—Splittings "show significant variations with structure, but these variations are by no means worse than variations in $^3J_{HH}$ previously used to assign configuration [in compounds of general structure RCH=CHR']."[53]

Geminal Couplings and Additivity Effects in Olefins

It was recognized at an early stage that chemical shift assignments could be made by means of additivity effects.[54] Specifically, the effects of substituents upon the chemical shifts of carbon in monosubstituted compounds could be algebraically added to predict the chemical shifts in compounds with two or more of these substituents. One had to be careful to use the same system (and, for example, not use in naphthalenes the additivity parameters developed for substituted benzenes) and to be aware of steric effects (for example, that a sterically perturbed carbon generally experiences an upfield shift of its ^{13}C NMR signal), but additivity parameters have proved to be probably the most powerful tool in making chemical shift assignments.

85   **86**   **87**

Additivity parameters also have been used to predict J_{CH} values.[49,55] Compounds **85–87** are the model systems to supply the additivity parameters for $^2J_{CH}$ values.[49] An example of the utilization of additivity parameters may be afforded by *cis*-dichloroethylene (**88**).[49] Consider the predicted value of the $^2J_{CH}$ value observed as 15.4 Hz. This coupling may be viewed as an algebraic sum of the two parameters ΔJ identified in **88a** and **88b**. The effect of a chlorine

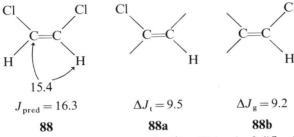

$$J_{pred} = 16.3 \qquad \Delta J_t = 9.5 \qquad \Delta J_g = 9.2$$

88   **88a**   **88b**

substituent trans to the coupled proton is $\Delta J = [7.1 - (-2.4)] = 9.5$ Hz, i.e., the chlorine substituent causes an algebraic increase of 9.5 Hz (**88a**). The effect of the chlorine substituent geminal to the proton is $\Delta J = [6.8 - (-2.4)] = 9.2$ Hz, i.e., the substituent causes an algebraic increase of 9.2 Hz (**88b**). The overall predicted value for **88** would be $J_{pred} = -2.4 + 9.5 + 9.2 = 16.3$ Hz. The agreement for **88** of $J_{obs} = 15.4$ and $J_{pred} = 16.3$ is as striking as is that for chemical shift additives. The algebraic additivity is validated by means of *trans*-dichloroethylene (**89**). For **89a** $\Delta J = [6.8 - (-2.4)] = 9.2$ Hz, and for **89b** $\Delta J = [-8.3 - (-2.4)] = -5.9$. The predicted value is $J_{pred} = -2.4 + 9.2 - 5.9 = 0.9$,

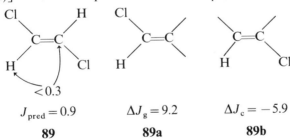

$$J_{pred} = 0.9 \qquad \Delta J_g = 9.2 \qquad \Delta J_c = -5.9$$

89   **89a**   **89b**

i.e., close to zero just as $J_{obs} < 0.3$. The impressive agreement between J_{obs} and J_{pred} continues through a series of di- and trisubstituted bromo- and chloro-ethylenes.[49] Additionally, for a series of olefins of wide functional variety (Table 2-8),[56] an extensive list of additivity parameters has been developed (Table 2-9).[56] These additivity parameters may be especially suitable for highly substituted olefins which lack not only vicinal H—H couplings but also easily extractable vicinal C—H couplings. An example is $PhClC=CHCHO$, whose observed $^2J_{CH}$ value of 7.2 Hz agrees with the calculated value for the Z geometry (+7.6 Hz) but not with the calculated value for the E geometry (−3.3 Hz).[56]

TABLE 2-8. GEMINAL CARBON–PROTON COUPLINGS IN OLEFINS[a] $\underset{Z}{\overset{Y}{>}}C_1{=}C_2\underset{X}{\overset{H}{<}}$ WHERE X, Y, Z

= SUBSTITUENTS AND/OR H

Compound			$^2J_{CH}$			
X	Y	Z	$J_{C_1\text{-}H}$	$J_{C_1\text{-}X}$	$J_{C_2\text{-}Y}$	$J_{C_2\text{-}Z}$
H	H	H	−2.4	−2.4	−2.4	−2.4
Cl	H	H	+6.8		+7.1	−8.3
Br	H	H	+5.8		+7.5	−8.5
I	H	H	+4.0		+4.15	−7.8
OEt	H	H	(+)9.7		(+)5.3	(−)5.3
OCH=CH$_2$	H	H	(+)9.75			
OCOCH$_3$	H	H	+9.65		+7.6	−7.9
NDCOCH$_3$	H	H			+3.5	(−)5.3
Pyrrolidinyl	H	H	(+)3.5		(+)3.5	(−)4.7
SiCl$_3$	H	H	−6.85		−2.5	−0.8
Si(CH=CH$_2$)$_3$	H	H	−7.0			
CH$_3$	H	H	+0.35		−1.15	−2.60
CH=CH$_2$	H	H	−0.05			
Ph	H	H	−1.0		0.0	−4.5
C≡CH	H	H	+8.8(?)		−0.3	−3.7
CN	H	H	−3.67		+0.29	−4.42
CHO	H	H	+0.25		−0.6	−3.4
COCH$_3$	H	H	−0.25		−1.1	−3.2
COOH	H	H	−0.6		−0.2	−4.55
COOCH$_3$	H	H	−0.6			
Cl	Cl	H	+0.8			+0.8
Br	Br	H	−0.4			−0.4
I	I	H	−1.4			−1.4
Cl	CH$_3$	H	(+)6.2			(−)6.2
Cl	COOH	H	(+)3.8			−6.1
Cl	CON(CH$_3$)$_2$	H				−6.8
Br	CH$_3$	H	(+)5.7			(−)6.2
Br	COOH	H	(+)3.1			(−)6.4
I	COOH	H	(+)0.9			−5.9
OMe	COCH$_3$	H	(+)8.0			(−)3.5
OEt	CH$_3$	H	(+)10.0			
CN	CN	H	−5.9			−5.9
CN	CH$_3$	H	−4.2			−1.9
CO$_2$Et	CO$_2$Et	H	−2.8			−2.8
CO$_2$Et	Ph	H				−3.3
COOH	CH$_3$	H	−0.5			−1.7
CHO	CH$_3$	H				−0.8
Cl	H	Cl	+16.0		+16.0	

TABLE 2-8. (continued)

Compound			$^2J_{CH}$			
X	Y	Z	$J_{C_1\text{-}H}$	$J_{C_1\text{-}X}$	$J_{C_2\text{-}Y}$	$J_{C_2\text{-}Z}$
Br	H	Br	+14.7		+14.7	
I	H	I	+11.0		+11.0	
Cl	H	CH$_3$	(+)7.4		(+)10.5	
Cl	H	COOH	(+)6.6		(+)7.7	
Br	H	CH$_3$	(+)6.8		(+)10.8	
Br	H	COOH	(+)6.2		(+)7.3	
I	H	COOH	(+)4.1		(+)5.4	
OMe	H	C≡CH	(+)12.0		(+)3.7	
OEt	H	CH$_3$	(+)8.9			
CN	H	CH$_3$	−2.4		+3.3	
COOH	H	CH$_3$	−0.5		+0.2	
H	Cl	Cl	0.6	0.6		
H	Br	Br	(+)1.3	(+)1.3		
H	CN	Cl	+3.8	−6.3		
H	COOH	Cl	(+)3.7	(−)6.0		
H	Ph	Cl	(+)5.6	(−)6.3		
H	CH$_3$	Cl	(+)7.0	(−)7.0		
H	COOH	Br	(+)2.7	(−)5.9		
H	CH$_3$	Br	(+)7.0	(−)7.0		
H	CN	OAc	(+)3.9	(−)6.3		
H	CH$_3$	OAc	(+)6.8	(−)6.8		
H	CH$_3$	CN	1.0	(−)3.1		
H	CH$_3$	COOH	0.5	(−)3.1		
H	CH$_3$	CHO	1.0	(−)2.2		
Cl	Cl	Cl	(+)8.9			
Br	Br	Br	(+)8.0			
Cl	Cl	COOH	(+)1.0			
Cl	COOH	Cl	(+)11.7			
Br	Br	COOH	1.0			
Br	COOH	Br	(+)10.5			
I	I	COOH	1.0			
I	I	COOCH$_3$	1.0			
I	COOCH$_3$	I	(+)5.8			
COOH	Cl	CH$_3$	(−)6.7			
COOH	CH$_3$	Cl	(+)6.4			
Ph	CHO	Cl	(+)6.1			
OEt	COOH	CN	(+)10.1			
COOH	CH$_3$	CH$_3$	1.3			
Ph	CH$_3$	CH$_3$	1.9			

[a]Reference 56. Signs in parentheses are determined from calculated J values, using additivity parameters of Table 2-9. These calculated J values depart more noticeably from the empirical values for the cis-disubstituted and trisubstituted olefins, but never by more than 2.5 Hz. (Reproduced with permission).

TABLE 2-9. ADDITIVITY PARAMETERS FOR $^2J_{CH}$ VALUES OF OLEFINS[a]

Substituent	ΔJ_g	ΔJ_c	ΔJ_t
F	+17	−11	+10
Cl	+9.2	−5.9	+9.5
Br	+8.2	−6.1	+9.1
I	+6.4	−5.4	+6.5
OEt	+12.1	−2.9	+7.7
OCH=CH$_2$	+12.1		
OCOCH$_3$	+12.0	−5.5	+10.0
NDCOCH$_3$		−2.9	+5.9
Pyrrolidinyl	+5.9	−2.3	+5.9
SiCl$_3$	−4.4	+1.6	−0.1
Si(CH=CH$_2$)$_3$	−4.6		
CH$_3$	+2.7	−0.2	+1.2
CH=CH$_2$	+2.3		
Ph	+1.4	−2.1	+2.4
C≡CH	−1.6	−1.3	+2.1
CN	−1.3	−2.0	+2.7
CHO	+2.6	−1.0	+1.8
COCH$_3$	+1.8	−2.2	+2.2
COOH	+1.8	−2.2	+2.2
COOR	+1.8		
CON(CH$_3$)$_2$	+1.5		

[a]Reference 56. The values ΔJ_g, ΔJ_c, and ΔJ_t are the increments added to the basic $^2J_{CH}$ value of −2.4 Hz (for ethylene) for substituents, respectively, geminal, cis, and trans to the proton of the coupling path C=C—H. (Reproduced with permission).

Aromatic Couplings

With aromatic compounds it is possible to observe a variety of C—H couplings with the carbon either as a substituent (**90**) or as internally incorporated in the aromatic nucleus (**91**). For the couplings in **90**, trends are observed that reflect analogous behavior with proton–proton couplings. For

90   **91**

example, the o, m, and p couplings in **92** are all positive and attenuate with coupling distance, just as the proton–proton couplings in **93**.[2] Vicinal

carbon–proton couplings also appear to depend upon π-bond order, just as proton–proton couplings—for example, the C—H coupling in **94** is 4.7 Hz, the CH_3—C_1—C_2—H and H—C_1—C_2—CH_3 couplings in **95** and **96** (with a

higher π-bond order) are larger at 5.1–5.2 Hz, and the CH_3—C_2—C_3—H coupling in **96** (with lower π-bond order) is smaller at 4.1 Hz. Unlike vicinal couplings in olefins, the analysis of C—H couplings in aromatic compounds is not straightforward (owing to a greater number of protons) and necessarily involves the analysis of complex spectra or the synthesis of specifically deuterated compounds. Therefore, unfortunately few data exist for external vicinal couplings in aromatic compounds, and it is not known whether electronegativity effects of substituents are the same as for proton–proton couplings.

For internal carbon–proton couplings, the parent molecule benzene has been studied in three independent studies[55,57,58] and the average values are shown in **97**. It is difficult to see immediately how a proton–proton model exists for the couplings in **97** (because a proton cannot be multiply bonded), but ethylene **98** has been proposed[58] and works surprisingly well. The observed J_{HH} values in **98** are shown as $^2J_{HH} = +2.5$ Hz and trans-$^3J_{HH} = +19.1$ Hz.[27] Using the relationship $J_{CH} = 0.4 J_{HH}$ for an sp^2-hybridized carbon,[16] the calculated values are $^2J_{calc} = 0.4(+2.5) = +1.0$ and $^3J_{calc} = 0.4(+19.1) = +7.6$, in remarkable agreement with the observed values in **97**. For biphenylene,[59] naphthalene,[60]

97   **98**

π-benzenechromium tricarbonyl,[61] and π-tropyliumchromium tricarbonyl cation[61] the $^2J_{CH}$, $^3J_{CH}$, and $^4J_{CH}$ values exhibit comparable magnitudes as the respective values of benzene (**97**) and may vary in accordance with C—C bond lengths and π-bond order.[59]

Electronegative substituent effects for couplings in **97** have been well characterized for halogens (Cl, Br, I).[55] When the halogen is substituted onto the coupling carbon (**99**), the increments to the o, m, and p couplings are negative, positive, and negative, respectively (-4.5, $+3.5$, -0.8), in accord with the "alternating substituent effect"[28] proposed for aliphatics. The negative increment to $^2J_{CH}$ in **99** also parallels that observed for $^2J_{CH}$ in olefins with a cis electronegative substituent[49] (**100**; see Geminal Couplings and Additivity Effects in Olefins, this chapter) and the positive increment to $^3J_{CH}$ in **99** reflects similar behavior for trans-$^3J_{CH}$ couplings[52] (**101**; see Olefinic Couplings, this chapter).

99   **100**   **101**

When the halogen substituent is further removed from the coupling (**102–105**), the effect of the halogen is less, as expected. When the halogen is substituted internally to the coupling route (**106, 107**), the effect is small with the $^4J_{CH}$ values

102   **103**   **104**   **105**

(there are two coupling routes), but is a negative increment to $^3J_{CH}$ in **106**, in accord with the observation that internally substituted olefins (**108**) experience a decrease as occurs with trans-$^3J_{CH}$[52] (see Olefinic Couplings, this chapter).

^{13}C Cl +5.0 H -1.2 **106** ^{13}C Cl -1.0 **107** ^{13}C Cl H **108**

These observations made for halogenated benzenes **99, 102–107**, have been generalized for the substituents F, Cl, Br, I, NO_2, NH_2, CHO, CN, OH, OCH_3, and $SiMe_3$.[62] Representative examples are as follows. Compounds **109–111**

^{13}C F H -4.9 H +11.0 H -1.7 **109** ^{13}C CH_3 H +0.5 H +7.6 H -1.4 **110** ^{13}C $SiMe_3$ H +4.2 H +6.3 H -1.1 **111**

illustrate a spectrum of electronegativity for the substituent directly bonded to the coupling carbon (corresponding to **99** above): $^2J_{CH}$ and $^4J_{CH}$ increase, and $^3J_{CH}$ decreases, as the electronegativity of the substituent decreases. In compounds **112–114** (corresponding to **106** above) the effect of an internally

^{13}C F H +4.1 **112** ^{13}C CH_3 H +6.6 **113** ^{13}C $SiMe_3$ H +8.6 **114**

substituted substituent on $^3J_{CH}$ is portrayed: As the electronegativity of the substituent decreases, $^3J_{CH}$ increases. For other placements of the substituent (corresponding to **102–105, 107** above), a much smaller change in J_{CH} is observed as the electronegativity of the substituent changes. All of these observations are in accord with anticipated trends, as just discussed for chlorobenzene.

Carbon–proton couplings to protons external to the aromatic ring are shown in **115** and **116**.[62] The J_{CH} values in **115** compare favorably with the equivalent couplings in propene **117**.

Structure Elucidation and Chemical Shift Assignments

The basic premise outlined in **92** and **97**—that $|{}^3J_{CH}| > |{}^2J_{CH}| \sim |{}^4J_{CH}|$, whether the carbon is internal or external to the aromatic ring—has been exploited in numerous structure elucidation and carbon chemical shift studies. In a straightforward application of this generalization, it was shown that dimethylphenols could be analyzed and identified by observing the coupling of the methyl carbon with the aromatic protons[63] [the methyl carbon was split into a quartet by the directly bonded protons (${}^1J_{CH}$), and the coupling in question was observed in each of four quartet signals]. Compounds **118–122** can be differentiated by these ${}^3J_{CH}$ couplings: **118** exhibits a singlet and a doublet

(the C_3 methyl is ortho to a proton, but the C_2 methyl is ortho to none), **119** shows one doublet, **120** two doublets, **121** one triplet, and **122** a doublet and a triplet. Similar analyses can help in the structure elucidation of methoxy-, amino-, and hydroxypyridines;[64] cyanopyridines;[65] and coumarin derivatives.[66] In studies of salicylaldehyde[67] and acylphloroglucinols,[68] the nature of the hydrogen bonding of phenolic OH groups was elucidated by means of C—H couplings through the oxygen:[68]

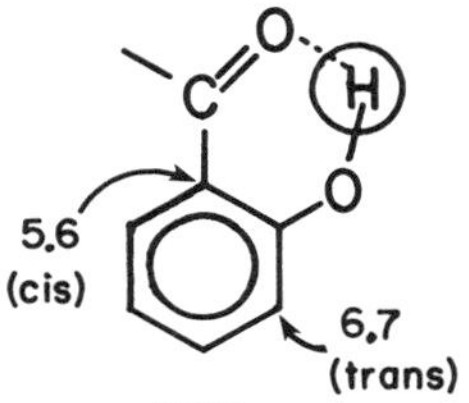

123

When a hydroxyl group was frozen into a particular conformation (123) by hydrogen bonding, the cis and trans $^3J_{CCOH}$ values were found to be 5.6 and 6.7 Hz, respectively.

A technique utilizing J_{CD} couplings has proved very powerful in making chemical shift assignments in aromatic compounds.[69, 70] A specifically deuterated compound, such as 1-methyl-4-deuterionaphthalene (124), can have carbon chemical shift assignments made in proton-decoupled (but deuterium-coupled) experiments by recognizing the following: (1) directly bonded carbons are split by the deuterium by a large amount (C-4 in 124); (2) $^3J_{CD}$ couplings are observable, leading in 124 to splitting in C-2 and C-5 (and coupling to C-5 will be smaller, being cis); (3) $^2J_{CD}$ couplings are small and are generally unobservable, but deuterium isotope effects are experienced, leading in 124 to an upfield shift of C-3 and C-4a relative to undeuterated 1-methylnaphthalene. Thus, carbons C-2, C-3, C-4, C-4a, and C-5 are unequivocally assigned. Similarly, assignments for compounds such as 125–128 can be simplified by the same arguments.

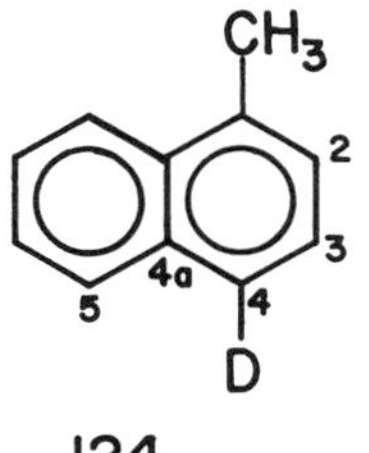

124

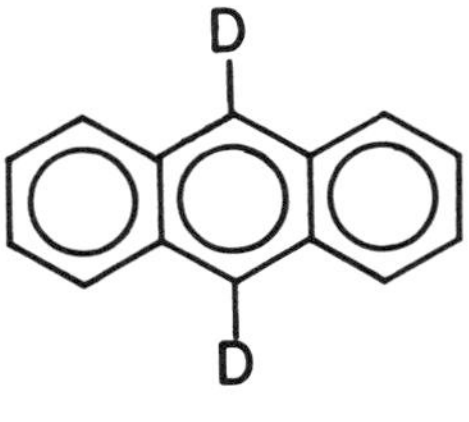

125

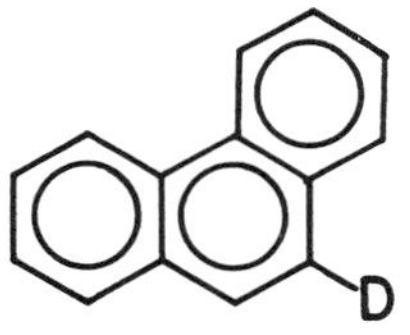

126

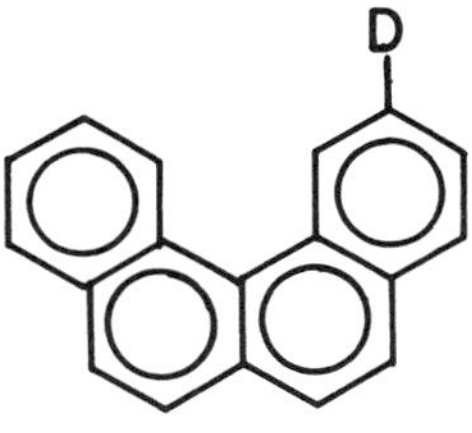

127

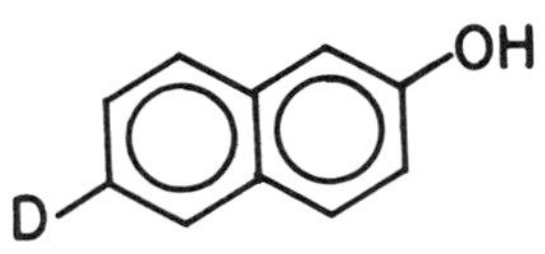

128

Additivity Effects

The carbon–proton coupling constants determined for monohalobenzenes (**99, 102–107**)[55] have been used to determine additivity parameters, just as for olefins (see Geminal Couplings and Additivity Effects in Olefins, this chapter). These additivity effects were then used to calculate C—H values for dihalobenzenes (**129–131**). These calculated J_{CH} values were in remarkable agreement

with the observed J_{CH} values and show that in compounds of rigid geometry (where conformational effects are of no concern) a set of additivity values can be developed to predict J_{CH} values in molecules of multifunctionality. An example of o-dichlorobenzene (**129**) is given here to illustrate how the method works. From the observed J_{CH} values for benzene (**97**) and the change observed for chlorobenzene (**99, 102–107**) the following parameters are developed (by subtracting the values in **97** from the respective values in **99, 102–107**):

The J_{calc} values for o-dichlorobenzene[55] then are determined by adding the two parameters to benzene. For example, the first J_{calc} value in **138** ($^2J_{CH}$) is calculated by adding to $^2J_{CH}$ from **97** ($+1.1$) the additivity effect from **133** (0.3 Hz) and the additivity effect from **134** (0.5 Hz). The agreement between J_{calc} and J_{obs} in **138–140** is good (maximum error $= 0.6$ Hz) and attests to the additivity of coupling contributions to an overall observed value of J_{CH}.

$$J_{calc} = (1.1 + 0.3 + 0.5) = 1.9 \qquad J_{obs} = 1.9$$
$$J_{calc} = (7.6 + 0.2 + 0.6) = 8.4 \qquad J_{obs} = 8.4$$
$$J_{calc} = (-1.2 + 0.0 + 0.3) = -0.9 \qquad J_{obs} = -1.2$$

138

$$J_{calc} = (1.1 - 0.8 - 0.2) = 0.1$$
$$J_{obs} = 0.0$$

$$J_{calc} = (1.1 + 0.5 - 0.2) = 1.4 \qquad J_{obs} = 1.1$$
$$J_{calc} = (7.6 + 0.6 - 0.2) = 8.0 \qquad J_{obs} = 8.6$$

139

$$J_{calc} = (1.1 - 4.5 + 0.3) = -3.1 \qquad J_{obs} = 3.5$$
$$J_{calc} = (7.6 + 3.5 + 0.2) = 11.3 \qquad J_{obs} = 11.6$$

$$J_{calc} = (7.6 + 3.5 - 2.6) = 8.5$$
$$J_{calc} = (-1.2 - 0.8 + 0.0) = -2.0 \qquad J_{obs} = -1.8$$
$$J_{obs} = 7.9$$

140

This method of predicting J_{CH} values from additivity principles has been extended to trisubstituted benzenes and to additional functional groups [Cl, NH_2, NEt_2, $N(iPr)_2$, $N(C_2H_4)_2O$] for three-bonded couplings.[71] For $^3J_{CH}$ couplings five different geometries are possible with respect to the coupling framework and to the substituent (**141–145**). The additivity parameters developed for each of these situations are denoted by Δ_1, Δ_2, Δ_3, Δ_4, and Δ_d, respectively, which are listed in Table 2-10. An example of how this method performs for trisubstituted benzene compounds is 2,4-dichloroaniline (**146**) (note that the authors used $^3J_{CH} = 7.4$ Hz as the value in parent benzene). Agreement between J_{obs} and J_{calc} values is sufficient to suggest that additivity parameters

TABLE 2-10. ADDITIVITY PARAMETERS FOR CALCULATING $^3J_{CH}$ VALUES IN BENZENE DERIVATIVES[a]

	Substituent				
Parameter[b]	Cl	NH$_2$	NEt$_2$	N(iPr)$_2$	N(C$_2$H$_4$)$_2$O
Δ_1	−2.24	−2.23	−1.47	−1.90	−2.03
Δ_2	+0.97	0.29	0.01	0.84	0.70
Δ_3	−0.17	−0.22	0.24	1.55	−0.38
Δ_4	−0.16	−0.24	0.35	−1.55	0.05
Δ_d	3.00	1.34	1.10	0.70	1.54

[a]Reproduced with permission from Wiley Heydon Ltd.[71]
[b]Structure of parameter defined by **141–145**. These parameters are added to the $^3J_{CH}$ value of 7.4 in benzene.

may be used in structure elucidation to decide unambiguously the placement of groups about a polysubstituted benzene. Hence, additivity parameters for J_{CH} values may complement the same method for chemical shifts, provided a sufficient number of additivity values are determined.

$$J_{C_1-H_3} = 7.4 + \Delta_d(NH_2) + \Delta_1(Cl) + \Delta_4(Cl)$$
$$= 7.4 + 1.34 - 2.24 - 0.16$$
$$= 6.3 \ J_{obs} = 6.2$$

$$J_{C_1-H_5} = 7.4 + \Delta_d(NH_2) + \Delta_3(Cl) + \Delta_4(Cl)$$
$$= 7.4 + 1.34 - 0.17 - 0.16$$
$$= 8.4 \ J_{obs} = 8.8$$

$$J_{C_3-H_5} = 7.4 + \Delta_2(NH_2) + \Delta_1(Cl) + \Delta_3(Cl)$$
$$= 7.4 + 0.29 - 2.24 - 0.17$$
$$= 5.3 \; J_{obs} = 5.7$$

$$J_{C_2-H_6} = 7.4 + \Delta_1(NH_2) + \Delta_d(Cl) + \Delta_2(Cl)$$
$$= 7.4 - 2.23 + 3.00 + 0.97$$
$$= 9.1 \; J_{obs} = 8.8$$

$$J_{C_4-H_6} = 7.4 + \Delta_4(NH_2) + \Delta_d(Cl) + \Delta_2(Cl)$$
$$= 7.4 + 0.24 + 3.00 + 0.97$$
$$= 11.6 \; J_{obs} = 12.0$$

$$J_{C_5-H_3} = 7.4 + \Delta_2(NH_2) + \Delta_1(Cl) + \Delta_4(Cl)$$
$$= 7.4 + 0.29 - 2.24 - 0.16$$
$$= 5.3 \; J_{obs} = 5.9$$

The Correlation of J_{CH} and J_{HH}: The J_{CH}/J_{HH} Ratio

As noted earlier (General Aspects of Aliphatic C—H Couplings, this chapter), the C—H couplings in **38** correlate with the H—H couplings in **39**, with the relationship observed[29]

$$^3J_{CH} = 1.2 \times {}^3J_{HH} - 3.5 \tag{12}$$

However, it is bothersome that the intercept is nonzero; at low values of $^3J_{CH}$ or $^3J_{HH}$ (e.g., when the dihedral angle is near 90°), one would predict from eq. (12) opposite signs for $^3J_{CH}$ and $^3J_{HH}$ and further would expect $^3J_{CH} = -3.5$ Hz when

CH₃   CH₃
| |
H₃C—C—X   H₃C—C—X
| |
CH₃   H

38   39

$^3J_{HH}$ is 0 Hz—an extremely improbable situation. Indeed, the validity of the Karplus-like relationship for $^3J_{CH}$ couplings (see The Karplus Dihedral Angle Relationship, this chapter) shows $^3J_{CH}$ is near 0 Hz for a dihedral angle of 90°. The difficulty with using an expression such as eq. (12) is that, because it was developed with no points near $J_{CH} = 0$, $J_{HH} = 0$, it is quite sensitive to the vicissitudes of ^3J at higher values. Instead, perhaps a better predictive tool is the ratio J_{CH}/J_{HH}, which varies from 0.62 to 0.71 for compounds **38** and **39**. A compound that appears to support the $^3J_{CH}/^3J_{HH}$ idea is **7** (taken from Table 2-1), whose trans-$^3J_{CH}$ value (2.51 Hz) compared with the trans-$^3J_{HH}$ value (3.70 Hz) of **16** gives a ratio of $^3J_{CH}/^3J_{HH} = 0.68$, in agreement with the range for

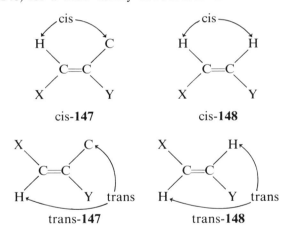

compounds **38** and **39**, even though these ^3J values are much lower than those of compounds **38** and **39** (the cis-$^3J_{CH}$ and cis-$^3J_{HH}$ values of **7** and **16** give a lower ratio of 0.50, but *cis*-carbonyl compounds are susceptible to through-space negative contributions: see Chapter 3, $^3J_{CC}$ vs. Orientation of a Terminal Substituent). A number of studies with olefins, also, have shown that cis and trans three-bonded C—H couplings (in **147**) correlate with analogous H—H couplings (in **148**) for a wide variety of substituents X and Y. Most of these

studies involve the coupling carbon as a methyl group, but some work includes functionalized carbon as a carboxyl group (see the Correlation of Carbon–Proton and Proton–Proton Couplings, this chapter). An extensive amount of data shows that in general the ratio $^3J_{CH}/^3J_{HH}$ for geometrically equivalent systems lies in the range 0.5–0.65.

In a series of isopropenyl derivatives **149** both cis and trans $^3J_{CH}$ values were determined and compared with geometrically equivalent $^3J_{HH}$ values of **150**.[72] The substituent X was F, OAc, OCH$_3$, Cl, Br, COCl, CO$_2$Me, CN, CH$_3$, Ph, Ph-p, C(Me)=CH$_2$, or CMe$_3$. The cis $^3J_{CH}$ values of **149** varied from 2.5 to 7.1 Hz as the electronegativity of X decreased through the series listed above, and the cis $^3J_{HH}$ values of **150** varied from 4.7 to 11.5 Hz correspondingly. For the trans couplings of **149** $^3J_{CH}$ varied from 7.0 to 11.5 Hz as the corresponding $^3J_{HH}$ values of **150** varied from 12.7 to 18.6 Hz. The plot of $^3J_{CH}$ vs. $^3J_{HH}$ gives a slope of 0.62 with a correlation coefficient $R = 0.974$ (intercept $= -0.24$).

In another study a wider variety of olefins with general structures **149**, **151**, and **152** was examined.[73] The $^3J_{CH}$ values obtained from these compounds (14 cases) may be compared with the geometrically equivalent $^3J_{HH}$ values (again obtained by replacing the CH$_3$ group with a proton) to give:

$$^3J_{CH} = 0.46 \times {^3J_{HH}} + 1.58 \qquad r = 0.956 \qquad (13)$$

The slope of 0.46 is misleading, because (as for the aliphatic case) the intercept in this case is not near the origin, owing to a few irregular points and to the lack of points near the origin. Instead, a better indicator of the true relationship of J_{CH} and J_{HH} is to determine the J_{CH}/J_{HH} ratio. Indeed, this ratio is $J_{CH}/J_{HH} = 0.60$ with a standard deviation of 0.06.

$$\frac{J_{CH}}{J_{HH}} = \frac{3.2}{5.05} = 0.63$$

$$\frac{J_{CH}}{J_{HH}} = \frac{6.1}{9.65} = 0.63$$

$$\frac{J_{CH}}{J_{HH}} = \frac{1.6}{3.0} = 0.53$$

Another wide range of olefins[52] primarily involving the general structure **149** but also including the more unusual examples **153–154** (23 cases) gives a slope of 0.62 with $R = 0.989$ (intercept $= -0.14$). These examples **153–154** attest to the remarkable similarity between $^3J_{CH}$ and $^3J_{HH}$ olefinic couplings. In compounds **153** and **154** both $^3J_{CH}$ and $^3J_{HH}$ reflect the π-bond order of the C—C bond between the coupling nuclei: the ratio $J_{CH}/J_{HH} = 0.63$ whether this C—C bond is the C_1—C_2 linkage or the C_2—C_3 linkage of the diene. Likewise, the flaring of the two hydrogens in **156** which causes a decrease in $^3J_{HH}$ is mirrored in **155**, whose $^3J_{CH}$ is similarly reduced in value,[52] and whose $J_{CH}/J_{HH} = 0.53$.

It was observed from compounds **153** and **154** that both $^3J_{CH}$ and $^3J_{HH}$ depend upon the π-bond order of the intervening C—C bond, and that the ratio $^3J_{CH}/^3J_{HH}$ is constant at 0.63. Aromatic compounds also reflect this behavior precisely. In **157** the two $^3J_{CH}$ couplings are directly dependent upon the π-bond order of the C—C bond and parallel the analogous $^3J_{HH}$ couplings of **158**.[52] The two J_{CH}/J_{HH} ratios from **157** and **158** are 0.60 and 0.61. In benzene derivatives **159** and **160** the trend continues: the $^3J_{CH}$ value in **159** is intermediate between the two $^3J_{CH}$ values of **157**, just as the $^3J_{HH}$ value in **160** is intermediate between the two $^3J_{HH}$ values in **158**, all reflecting the relative π-bond order of the involved C—C bond. Further, the ratio J_{CH}/J_{HH} for **159/160** is 0.62. This similarity

$$\frac{J_{CH}}{J_{HH}} = \frac{4.1}{6.85} = 0.60$$

$$\frac{J_{CH}}{J_{HH}} = \frac{5.1}{8.28} = 0.61$$

$$\frac{J_{CH}}{J_{HH}} = \frac{4.7}{7.54} = 0.62$$

between $^3J_{CH}$ and $^3J_{HH}$ values in aromatics also has been noted in heteronuclear aromatic compounds (quinolines[74] and pyridines[74, 75]) and nonbenzenoid polynuclear hydrocarbons (azulenes and aceheptylenes).[76]

To compare geometrically equivalent $^3J_{CH}$ and $^3J_{HH}$ values, Table 2-11 lists the $^3J_{CH}$ values for aliphatic, olefinic, and aromatic systems (where the coupling carbon is a substituent) and includes the $^3J_{CH}/^3J_{HH}$ ratio. A remarkable consistency is to be noted in this ratio, with extremes of 0.50–0.85. The average value is 0.614, with a standard deviation of 0.065. Such reliability suggests this ratio could be used with confidence in structural elucidations.

TABLE 2-11. THE RATIO J_{CH}/J_{HH} IN GEOMETRICALLY EQUIVALENT SYSTEMS FOR VICINAL ALIPHATIC, OLEFINIC, AROMATIC, AND ACETYLENIC COUPLINGS[a] (ALSO INCLUDES MISCELLANEOUS ACETYLENIC J_{CH} VALUES)

Compound	$^3J_{CH}$	$^3J_{CH}/^3J_{HH}$	Reference
$(CH_3)_2CDCH_3$	5.27	0.71	29
$(CH_3)_2C(CH_3)CH_3$	4.62	0.68	29
$(CH_3)_2CICH_3$	4.45	0.66	29
$(CH_3)_2CBrCH_3$	4.22	0.64	29
$(CH_3)_2CClCH_3$	4.08	0.64	29
$(CH_3)_2CFCH_3$	3.81	0.62	29
1,2,3,4,7,7-Hexachloronorbornene- endo-2-carboxylic acid			
$^3J_{COOH,H_{cis}}$	4.54	0.50	2
$^3J_{COOH,H_{trans}}$	2.51	0.68	2
$H_2C=CFCH_3$	2.5, 7.0	0.53, 0.55	72
$H_2C=COAcCH_3$	3.4, 8.1	0.53, 0.58	72
$H_2C=CClCH_3$	4.2, 8.6	0.58, 0.58	72
$H_2C=CBrCH_3$	4.5, 9.0	0.62, 0.60	72
$H_2C=C(COCl)CH_3$	5.6, 9.6	0.55, 0.57	72
$H_2C=C(CO_2Me)CH_3$	6.0, 10.3	0.57, 0.60	72
$H_2C=C(CN)CH_3$	6.4, 10.1	0.54, 0.56	72
$H_2C=C(Ph)CH_3$	6.7, 11.1	0.58, 0.60	72
$H_2C=C(Ph—F—p)CH_3$	6.8, 11.2	0.62, 0.64	72
$H_2C=C(CMe=CH_2)CH_3$	7.1, 11.2	0.70, 0.65	72
$H_2C=C(CMe_3)CH_3$	7.1, 11.5	0.71, 0.68	72
$ClHC=CHCH_3$	5.6, 8.1	0.77, 0.55	73
$CNHC=CHCH_3$	6.4, 9.3	0.54, 0.52	73
$(HCO)HC=CHCH_3$	5.9	0.59	73
$(HOOC)HC=CHCH_3$	6.3	0.62	73
$(EtOOC)HC=CHCH_3$	6.3		73
$(CH_3OC)HC=CHCH_3$	6.6	0.61	73
$H_2C=C(CHMe_2)CH_3$	6.9, 11.3	0.66, 0.66	52
$H_2C=C(CH_2Ph)CH_3$	6.6, 10.7	0.66, 0.63	52
$H_2C=C(Ph)CH_3$	6.7, 11.1	0.60, 0.62	52
$H_2C=C(Ph—Me—p)CH_3$	6.8, 11.2	0.65, 0.64	52
$H_2C=C(C≡CH)CH_3$	6.3, 10.7	0.55, 0.62	52
$H_2C=C(CH_2OH)CH_3$	6.4, 11.2	0.61, 0.64	52
$H_2C=C(CH_2Cl)CH_3$	5.9, 10.3	0.58, 0.61	52
$H_2C=CBrCH_3$	4.6, 8.9	0.65, 0.60	52
$H_2C=C(CHO)CH_3$	5.7, 10.0	0.57, 0.57	52
$H_2C=C(COOH)CH_3$	5.7, 10.1	0.54, 0.59	52
$CH_3HC=CHCOOH$	6.8, 14.5	0.67, 0.85	2
$H_2C=CHCOOH$	7.6, 14.1	0.66, 0.74	2
1,2,3,5-Tetramethylbenzene	4.7	0.62	52
1-Methylnaphthalene, $^3J_{CH_3,H_2}$	5.2	0.63	52
2-Methylnaphthalene, $^3J_{CH_3,H_1}$	5.1	0.62	52
2-Methylnaphthalene, $^3J_{CH_3,H_3}$	4.1	0.60	52

TABLE 2-11. (continued)

Compound	$^3J_{CH}$	$^3J_{CH}/^3J_{HH}$	Reference
2-Methylquinoline, $^3J_{CH_3,H_3}$	2	0.5	74
3-Methylquinoline, $^3J_{CH_3,H_2}$	3	0.7	74
3-Methylquinoline, $^3J_{CH_3,H_4}$	5	0.6	74
4-Methylquinoline, $^3J_{CH_3,H_3}$	5	0.6	74
5-Methylquinoline, $^3J_{CH_3,H_6}$	5	0.6	74
6-Methylquinoline, $^3J_{CH_3,H_5}$	5	0.6	74
6-Methylquinoline, $^3J_{CH_3,H_7}$	4	0.6	74
7-Methylquinoline, $^3J_{CH_3,H_6}$	4	0.6	74
8-Methylquinoline, $^3J_{CH_3,H_7}$	5	0.6	74
2-Methylpyridine, $^3J_{CH_3,H_3}$	2	0.4	74, 75
3-Methylpyridine, $^3J_{CH_3,H_2}$	3	0.6	74, 75
3-Methylpyridine, J_{CH_3,H_4}	5	0.7	74, 75
4-Methylpyridine, J_{CH_3,H_3}	5	0.7	74, 75
Benzoic acid, J_{COOH,H_2}	4.1	0.53	2
Benzoic acid, J_{COOH,H_3} ($^4J_{CH}$)	1.11	0.79	2
4-Methylazulene, J_{CH_3,H_5}	6.0	0.65	76
5-Methylazulene, J_{CH_3,H_4}	5.8	0.58	76
5-Methylazulene, J_{CH_3,H_6}	5.5	0.60	76
6-Methylazulene, J_{CH_3,H_5}	5.9	0.59	76
3-Methylaceheptylene, J_{CH_3,H_4}	6.6	0.63	76
4-Methylaceheptylene, J_{CH_3,H_3}	6.3	0.60	76
5-Methylaceheptylene, J_{CH_3,H_6}	7.3	0.60	76
6-Methylaceheptylene, J_{CH_3,H_5}	7.9	0.65	76
$CH_3\!-\!C\!\equiv\!C\!-\!H$ ($^3J_{CH}$)	3.41	0.36	79
$CH_3\!-\!C\!\equiv\!C\!-\!H$ ($^3J_{CH}$)	4.65		79
$CH_3\!-\!C\!\equiv\!C\!-\!H$ ($^2J_{CH}$)	+50.11		79
$CH_3\!-\!C\!\equiv\!C\!-\!H$ ($^2J_{CH}$)	−10.43		79
$CH_3\!-\!C\!\equiv\!C\!-\!CH_3$($^3J_{CH}$)	+4.30		79
$CH_3\!-\!C\!\equiv\!C\!-\!CH_3$($^2J_{CH}$)	−10.34		79
$CH_3\!-\!C\!\equiv\!C\!-\!CH_3$($^4J_{CH}$)	+1.58	0.34	79
$H\!-\!C\!\equiv\!C\!-\!H$	+49.62		79

^{a}The geometrically equivalent J_{HH} value is derived from the compound where the coupling carbon has been replaced by a hydrogen.

Two-bonded carbon–proton couplings ($^2J_{CH}$ in **161**) do not correlate with two-bonded proton–proton couplings ($^2J_{HH}$ in **162**) as well as do three-bonded couplings. Substituent and hybridization effects have a more unpredictable

161   **162**

influence on $^2J_{CH}$ and $^2J_{HH}$. However, gross trends are compatible with the idea of overall correlation. For instance, the compounds **163–166** exemplify that for sp^3 systems (**163** and **164**) both $^2J_{CH}$ and $^2J_{HH}$ are negative, and $|^2J_{CH}| < |^2J_{HH}|$. Further, for carbonyl systems (**165** and **166**) both $^2J_{CH}$ and $^2J_{HH}$ are very large and again $|^2J_{CH}| < |^2J_{HH}|$.[77]

In intermediate cases, with other substituent effects and with olefins, correlations between $^2J_{CH}$ and $^2J_{HH}$ are more strained.[77] For 17 cases of sp^3 systems including **163** and **164**, a linear regression gave

$$^2J_{CH} = 0.551 \times {}^2J_{HH} + 4.929 \qquad (14)$$

but the slope is misleading because of an intercept far from the origin. Similarly, a linear regression of sp^2 systems including **165** and **166** (five cases) gave:

$$^2J_{CH} = 0.578 \times {}^2J_{HH} + 2.863 \qquad (15)$$

but again the intercept is worrisome. A better indicator of possible correlation, as before, would be the ratio J_{CH}/J_{HH}. For the 17 sp^3 cases[77] used to develop eq. (14), J_{CH}/J_{HH} is calculated to be 0.46, with a standard deviation of 1.40 (Table 2-12). For the five sp^2 cases[77] used to develop eq. (15), $J_{CH}/J_{HH} = 2.22$, with a standard deviation of 1.34 (Table 2-12). This $^2J_{CH}/^2J_{HH}$ ratio is quite different from the three-bonded ratio $^3J_{CH}/^3J_{HH}$ for the same compounds (see Table 2-11). This behavior of $^2J_{CH}$ values—that they are more positive than suggested from $^3J_{CH}/^3J_{HH}$ correlations—may reflect the flaring of the geminal angle for the C—H systems.[27] Furthermore, geminal couplings are notoriously sensitive to through-space interactions with α substituents;[27] and for π-systems the coupling nuclei are close to the nodal plane,[27] allowing great sensitivity to small changes in geometry.

Tables 2-11 and 2-12 therefore tell the story. For three-bonded couplings (Table 2-11) $^3J_{CH}$ and $^3J_{HH}$ values correlate well, with the average value of $^3J_{CH}/^3J_{HH} = 0.61$ with the total range 0.50–0.85. For two-bonded couplings, however (Table 2-12), there is a greater variance in the ratio $^2J_{CH}/^2J_{CH}$— generally this ratio is low (less than 0.5) for sp^3 systems, large (1.5–4) for olefin sp^2 systems, and intermediate (0.5–1.0) for carbonyl sp^2 systems. It is to be noted that Tables 2-11 and 2-12 include the coupling carbon both as —CH$_3$ and as —COOH for both sp^3 and sp^2 cases.

The tenth and eleventh entries in Table 2-12, 2-methyl-1,3-dioxane and 2-methyl-1,3-dioxolane (**167** and **168**, respectively) appear at first sight to be

TABLE 2-12. THE RATIO J_{CH}/J_{HH} IN GEOMETRICALLY EQUIVALENT SYSTEMS FOR GEMINAL ALIPHATIC AND OLEFINIC COUPLINGS[a]

Compound	$^2J_{CH}$	J_{CH}/J_{HH}	Reference
sp^3			
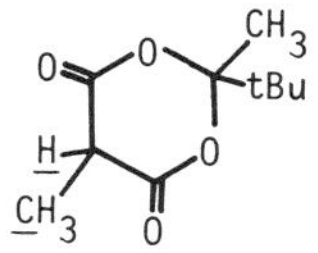	-6.2	0.29	77
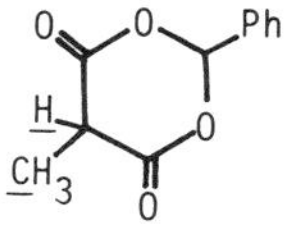	-5.4	0.29	77
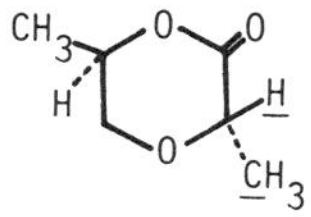	-4.3	0.24	77
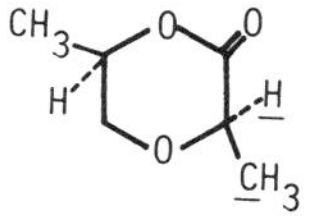	-4.2	0.24	77
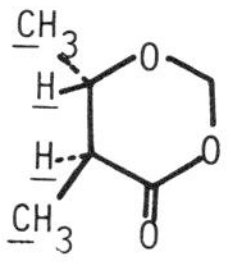	-1.3, -4.6	0.12, 0.26	77
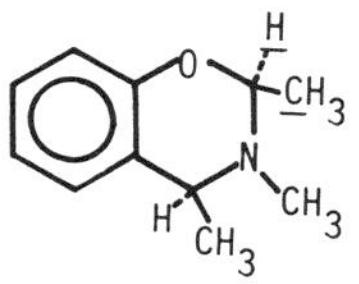	-4.6	0.28	77
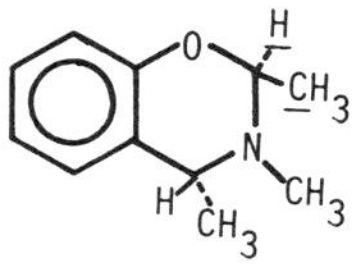	-4.7	0.28	77
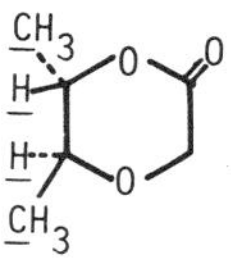	-1.6, -2.3	0.15, 0.18	77

TABLE 2-12. (continued)

Compound	$^2J_{CH}$	J_{CH}/J_{HH}	Reference
	−2.4, 2.3	0.20, −0.43	77
	+1.6	−0.26	77
	+6.0	6.0	77
	−4.6	0.45	2
	−6.17	0.49	2
	−6.89	0.55	2
sp²			
	+26.7	0.65	77
$CH_3\overset{\overset{\displaystyle O}{\|}}{C}$—H,2,4-dinitrophenylhydrazone	+9.1	0.78	77
	+3.9	2.30	77

TABLE 2-12. (continued)

Compound	$^2J_{CH}$	J_{CH}/J_{HH}	Reference
H_3C, H / $C=C$ / H, CHO	+3.5	3.5	77
H_3C, H / $C=C$ / H, CN	+0.9	3.9	77
$HOOC$, H / $C=CH_2$	+2.5	1.64	2
$HOOC$, CH_3 / $C=C$ / H, H	+3.1	1.44	2
$HOOC$, H / $C=C$ / H, CH_3	+3.4	1.56	2

[a] The geometrically equivalent J_{HH} value is derived from the compound where the coupling carbon has been replaced by a hydrogen.

departures from the overall trends. However, these compounds demonstrate that what happens when a change of sign of J occurs is perfectly in accord with previously established trends. First, it was previously observed that sp^3 $^2J_{CH}$ values are more positive than is expected from $^3J_{CH}$ trends, possibly because of the flaring of the CH_3—C—H angle. Second, the geminal proton–proton couplings in the H—H model compounds (**169** and **170**) themselves are more positive than normal, owing to lone-pair electron interactions of the oxygen atoms.[27] Third, the orientation of the lone-pair electrons with respect to the coupling nuclei has a very pronounced effect on the value of $^2J_{HH}$,[27] i.e., for the more nearly flat **170** the lone pair electrons eclipse the C—H bonds more effectively and cause a positive increment in $^2J_{HH}$. Thus, it is perfectly reasonable

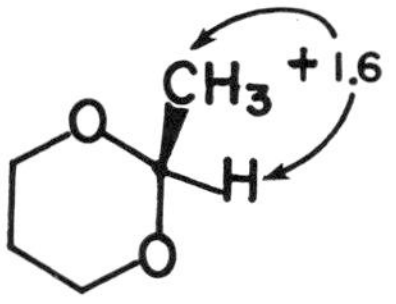

167 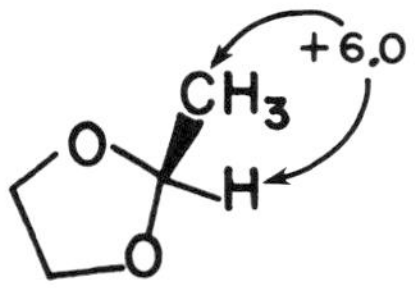   **168**

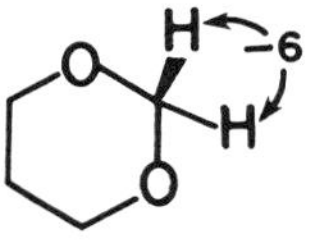

169 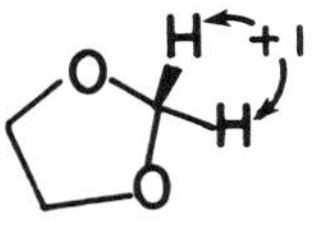   **170**

to expect that $^2J_{CH}$ for **168** is greater than $^2J_{CH}$ for **167**, and because $^2J_{HH}$ for **170** itself is positive it is hardly surprising that $^2J_{CH}$ for **167** and particularly $^2J_{CH}$ for **168** are positive.

Following the idea of the analogy between **97** and **98** (see Aromatic Couplings, this chapter), a correlation was noted between J_{CH} values in monosubstituted benzenes and J_{HH} values in olefins.[62] Structures **171–178** show the compounds related. Except for **171** and **172** (two-bonded couplings), the J_{CH}/J_{HH} ratio remains in the range observed generally for three-bonded couplings. Thus, the J_{CH}/J_{HH} ratio can be used even when the coupling carbon is not a substituent but instead is part of a C—C π system (because the coupling carbon in **171**, **173**, **175**, and **177** is not a substituent, these data are not included in Table 2-11).

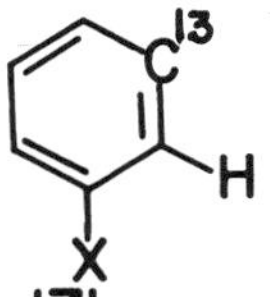

171   vs.   **172**

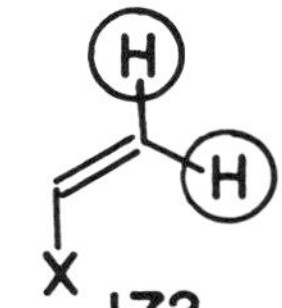

$$\frac{^2J_{CH}}{^2J_{HH}} = 0.8$$

Intercept $= 0.56$
$r = 0.92$

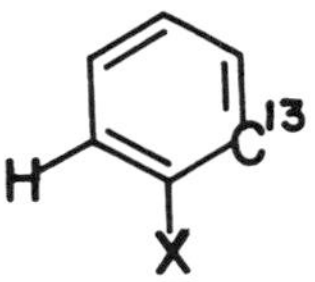

173   vs.   **174**

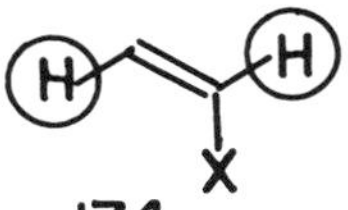

$$\frac{^3J_{CH}}{^3J_{HH}} = 0.575$$

Intercept $= -0.338$
$r = 0.96$

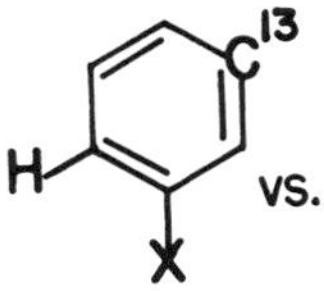

175   vs.   **176**

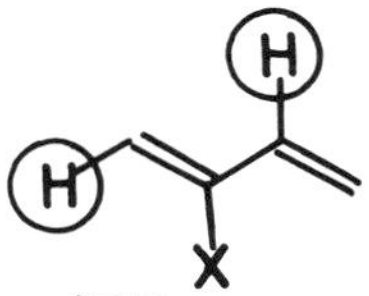

$$\frac{^4J_{CH}}{^4J_{HH}} = 0.608$$

Intercept $= -0.804$
$r = 0.99$

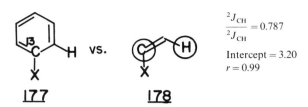

$$\frac{{}^2J_{CH}}{{}^2J_{CH}} = 0.787$$

Intercept $= 3.20$
$r = 0.99$

177   **178**

Along another vein, correlations also have been noted between ${}^2J_{CH}$ and ${}^3J_{HH}$ in the same series of olefinic molecules.[78] In the general structure **179**, the

$$\begin{array}{ccc} H_2 & & H_1 \\ & C_\beta{=}C_\alpha & \\ H_3 & & X \end{array}$$

$X = Cl, Br, I, COOH, OCH_2CH_3, SiEt_3,$ etc.

179

various geminal ${}^2J_{CH}$ values $(J_{C_\alpha\text{-}H_2}, J_{C_\alpha\text{-}H_3}, J_{C_\beta\text{-}H_1})$ were found to correlate with the sum of the vicinal ${}^3J_{CH}$ values (cis and trans), but of course this comparison is not between geometrically equivalent coupling paths. Instead, there is a dependence (either direct or inverse) of $J_{C_\alpha\text{-}H_2}, J_{C_\alpha\text{-}H_3}, J_{C_\beta\text{-}H_1}$, and J_{HCCH} on the electronegativity of the substituent.

In summary, J_{CH} and J_{HH} appear to correlate well for both two-bonded and three-bonded couplings. The excellent correlation for three-bonded couplings, for aliphatic, olefinic, and aromatic couplings, is particularly striking. For geminal couplings, one must be more careful because ${}^2J$ values may be positive or negative. One must be constantly aware that "a *general* simple correlation does not exist but it should be studied case by case by comparing the corresponding $J(HH)$ and $J(CH)$ for a particular type of compound and coupling pathway ...,"[73] but the same considerations appear to be operating for both H—H and C—C couplings—the Fermi contact mechanism, steric factors, and electronic contributions. Further, it should be understood that discussions in this section are limited to couplings in which the carbon is a substituent and for which a proton model is available. Finally, it should be noted that compounds in Tables 2-11 and 2-12 have the substituent mainly as —CH$_3$ and to a lesser extent —COOH; correlations for C—H couplings with the carbon embodied as a different substituent (such as —CH$_2$OH, —CHO, —CN) have not been done. However, one would expect to be able to use correlations for such substituents by using the anticipated substituent effects discussed earlier for aliphatic, olefinic, and aromatic C—H couplings. Thus, one can feel secure in using proton–proton couplings as an aid in estimating carbon–proton couplings in geometrically equivalent environments. The success with which a correlation has been found between J_{CH} and J_{HH} lends credence to the view that "empirical correlations appear to be more useful than theoretical calculations in reproducing the substituent effects upon J_{CH}."[62]

References

1. A number of papers discuss the importance of non-Fermi contact mechanisms to carbon–carbon couplings and generally agree that for directly bonded couplings ($^1J_{CC}$) these contributions are less significant than the Fermi contact mechanism and may be minimal, particularly for couplings in single bonds: Schulman, J. M.; Newton, M. D. *J. Am. Chem. Soc.*, **96**, 6295 (1974); Lazzeretii, P.; Taddei, F.; Zanasi, R. *J. Am. Chem. Soc.*, **98**, 7989 (1976); Lee, W. S.; Schulman, J. M. *J. Am. Chem. Soc.*, **101**, 3182 (1979). The question of non-Fermi contact contributions is also clouded by the question of the validity of using percentage s character to compute directly bonded C—C couplings: Pomerantz, M.; Liebman, J. F. *Tetrahedron Lett.*, 2385 (1975).
2. Marshall, J. L.; Seiwell, R. *Org. Magn. Reson.*, **8**, 419 (1976).
3. Bothner-By, A. A.; Castellano, S. M. In "Computer Programs for Chemistry," DeTar, D. F. ed.; Benjamin: New York, 1968; Vol. 1, p. 10.
4. Castellano, S. M.; Bothner-By, A. A. *J. Chem. Phys.*, **47**, 5443 (1967).
5. Karplus, M. *J. Am. Chem. Soc.*, **85**, 2870 (1963).
6. Murrell, J. N. In "Progress in NMR Spectroscopy," Emsley, J. W.; Feeney, J.; Sutcliffe, L. H. eds.; Pergamon Press: New York, 1970; Vol. 6, pp. 1–60.
7. Armour, E. A. G.; Stone, A. J. *Proc. R. Soc. London Ser. A*, **302**, 25 (1975).
8. Electron density should also be a consideration: electronegative terminal substituents should increase J_{CH}, and one would expect a smaller value of J_{CH} for hydrocarbons. For a methyl substituent in lieu of a carboxyl substituent, the ratio of J_{CH}/J_{HH} drops to 0.50, but the ratio is still greater than 0.25. This ratio of 0.5 is derived from J_{CH} values from alkanes and alkynes, Stothers, J. B. "Carbon-13 NMR Spectroscopy"; Academic Press: New York, 1972, pp. 349 and 360, and from alkenes, Anderson, J. E. *Tetrahedron Lett.*, 4079 (1975). Another consideration for the prediction of coupling constants is the mean excitation energy (ref. 6), but this term is exceedingly difficult to evaluate.
9. Paschal, J. W.; Rabideau, P. W. *J. Am. Chem. Soc.*, **96**, 272 (1974); Rabideau, P. W.; Paschal, J. W.; Patterson, L. E. *J. Am. Chem. Soc.*, **97**, 5700 (1975); Bennett, M. J.; Purdham, J. T.; Takada, S.; Masamune, S. *J. Am. Chem. Soc.*, **93**, 4063 (1971); Janadecek, R. J.; Simonsen, S. H. *J. Am. Chem. Soc.*, **91**, 6663 (1969).
10. Grossel, M. C.; Perkins, M. J. *J. Chem. Soc. Perkin Trans. 2*, 1544 (1975); Atkinson, D. J.; Perkins, M. J. *Tetrahedron Lett.*, 2335 (1969).
11. Brinkman, A. W.; Gordon, M.; Harvey, R. G.; Rabideau, P. W.; Stothers, J. B.; Ternay, A. L. Jr. *J. Am. Chem. Soc.*, **92**, 5912 (1970); Rabideau, P. W.; Paschal, J. W. *J. Am. Chem. Soc.*, **94**, 580 (1972).
12. Marshall, J. L.; Folsom, T. K. *J. Org. Chem.*, **36**, 2011 (1971); Marshall, J. L.; Ihrig, A. M.; Jenkins, P. N. *J. Org. Chem.*, **37**, 1863 (1972).
13. Marshall, J. L.; Faehl, L. G.; McDaniel, C. R. Jr.; Ledford, N. D. *J. Am. Chem. Soc.*, **99**, 321 (1977).
14. Karabatsos, G. J. *J. Am. Chem. Soc.*, **83**, 1230 (1961).
15. Karabatsos, G. J.; Graham, J. D.; Vane, F. J. *Phys. Chem.*, **65**, 1657 (1961).
16. Karabatsos, G. J.; Graham, J. D.; Vane, F. M. *J. Am. Chem. Soc.*, **84**, 37 (1962).
17. Karabatsos, G. J.; Orzech, C. E. Jr. *J. Am. Chem. Soc.*, **86**, 3574 (1964).
18. Karabatsos, G. J.; Orzech, C. E. Jr. *J. Am. Chem. Soc.*, **87**, 560 (1965).
19. Karabatsos, G. J.; Orzech, C. E. Jr.; Hsi, N. *J. Am. Chem. Soc.*, **88**, 1817 (1966).
20. Wasylishen, R. E.; Chum, K.; Bukata, J. *Org. Magn. Res.*, **9**, 473 (1977).
21. Wasylishen, R. E.; Schaefer, T. *Can. J. Chem.*, **52**, 3247 (1974).
22. Jameson, C. J.; Damasco, M. C. *Mol. Phys.*, **18**, 491 (1970).
23. Ewing, D. F. In "Annual Reports on NMR Spectroscopy," Mooney, E. F. ed.; Academic Press: New York, 1975; Vol. 6A, p. 389.
24. Lynden-Bell, R. M.; Sheppard, N. *Proc. R. Soc. London Ser. A*, **269**, 385 (1962).
25. Graham, D. M.; Holloway, C. E. *Can. J. Chem.*, **41**, 2114 (1963).
26. Jensen, H.; Schaumburg, K. *Mol. Phys.*, **22**, 1041 (1971).

27. Sternhell, S. *Quart. Rev.*, **23**, 236 (1969).
28. Wasylishen, R.; Schaefer, T. *Can. J. Chem.*, **51**, 961 (1973).
29. Forrest, T. P.; Sukumar, S. *Can. J. Chem.*, **55**, 3686 (1977).
30. Wasylishen, R.; Schaefer, T. *Can. J. Chem.*, **50**, 2710 (1972).
31. Lemieux, R. U.; Nagabhushan, T. L.; Paul, B. *Can. J. Chem.*, **50**, 773 (1972).
32. Delbaere, L. T. J.; James, M. N. G.; Lemieux, R. U. *J. Am. Chem. Soc.*, **95**, 7866 (1973).
33. Hamer, G. K.; Balza, F.; Cyr, N.; Perlin, A. S. *Can. J. Chem.*, **56**, 3109 (1978).
34. Kingsbury, C. A.; Jordan, M. E. *J. Chem. Soc. Perkin Trans. 2*, 364 (1977).
35. Schwarcz, J. A.; Perlin, A. S. *Can. J. Chem.*, **50**, 3667 (1972).
36. Lemieux, R. U.; Koto, S. *Tetrahedron*, **30**, 1933 (1974).
37. Spoormaker, T.; de Bie, M. J. A. *Rec. Trav. Chim.*, **99**, 15 (1980).
38. Cline, B. L.; Fagerness, P. E.; Panzica, R. P.; Townsend, L. B. *J. Chem. Soc. Perkin Trans. 2*, 1586 (1980).
39. Espersen, W. G.; Martin, R. B. *J. Phys. Chem.*, **80**, 741 (1976).
40. Feeney, J.; Hansen, P. E.; Roberts, G. C. K. *Chem. Commun.*, 465 (1974).
41. Rennekamp, M. E.; Kingsbury, C. A. *J. Org. Chem.*, **38**, 3959 (1973).
42. Hansen, P. E.; Feeney, J.; Roberts, G. C. K. *J. Magn. Reson.*, **17**, 249 (1975).
43. Barfield, M.; Grant, D. M. *J. Am. Chem. Soc.*, **85**, 1899 (1963).
44. Perlin, A. S.; Casu, B. *Tetrahedron Lett.*, 2921 (1969).
45. Perlin, A. S.; Cyr, N.; Ritchie, R. G. S.; Parfondry, A. *Carbohydr. Res.* **37**, C1 (1974).
46. Schwarcz, J. A.; Cyr, N.; Perlin, A. S. *Can. J. Chem.*, **53**, 1872 (1975).
47. Cyr, N.; Hamer, G. K.; Perlin, A. S. *Can. J. Chem.*, **56**, 297 (1978).
48. Lynden-Bell, R. M. *Mol. Phys.*, **6**, 537 (1963).
49. Weigert, F. J.; Roberts, J. D. *J. Phys. Chem.*, **73**, 449 (1969).
50. Dubs, R. V.; von Philipsborn, W. *Org. Magn. Reson.*, **12**, 326 (1979).
51. Anderson, J. E. *Tetrahedron Lett.*, 4079 (1975).
52. Vogeli, U.; von Philipsborn, W. *Org. Magn. Reson.*, **7**, 617 (1975).
53. Kingsbury, C. A.; Draney, D.; Sopchik, A.; Rissler, W.; Durham, D. *J. Org. Chem.*, **41**, 3863 (1976).
54. Grant, D. M.; Paul, E. G. *J. Am. Chem. Soc.*, **86**, 2984 (1964).
55. Tarpley, A. R. Jr.; Goldstein, J. H. *J. Phys. Chem.*, **76**, 515 (1972).
56. Vogeli, U.; Herz, D.; von Philipsborn, W. *Org. Magn. Reson.*, **13**, 200 (1980).
57. Günther, H.; Seel, H.; Günther, M.-E. *Org. Magn. Reson.*, **11**, 97 (1978).
58. Weigert, F. J.; Roberts, J. D. *J. Am. Chem. Soc.*, **89**, 2967 (1967).
59. Runsink, J.; Günther, H. *Org. Magn. Reson.*, **13**, 249 (1980).
60. Seel, H.; Aydin, R.; Günther, H. *Z. Naturforsch. Teil B*, **33**, 353 (1978).
61. Aydin, R.; Günther, H.; Runsink, J.; Schmickler, H.; Seel, H. *Org. Magn. Reson.*, **13**, 210 (1980).
62. Ernst, L.; Wray, V.; Chertkov, V. A.; Sergeyev, N. M. *J. Magn. Reson.*, **25**, 123 (1977).
63. Karliner, J.; Rodebaugh, R. *Tetrahedron Lett.*, 3783 (1975).
64. Takeuchi, Y.; Dennis, N. *Org. Magn. Reson.*, **7**, 244 (1975).
65. Takeuchi, Y.; Dennis, N. *J. Am. Chem. Soc.*, **96**, 3657 (1974).
66. Cussans, N. J.; Huckerby, T. N. *Tetrahedron*, **31**, 2719 (1975).
67. Äyräs, P.; Laatikainen, R.; Lötjönen, S. *Org. Magn. Reson.*, **13** 387 (1980).
68. Äyräs, P.; Widén, C.-J. *Org. Magn. Reson.*, **11**, 551 (1978).
69. Martin, R. H.; Moriau, J.; Defay, N. *Tetrahedron*, **30**, 179 (1974).
70. Kitching, W.; Bullpitt, M.; Doddrell, D.; Adcock, W. *Org. Magn. Reson.*, **6**, 289 (1974).
71. Nery, H.; Canet, D.; Azoui, B.; Lalloz, L.; Caubère, P. *Org. Magn. Reson.*, **10**, 240 (1977).
72. Douglas, A. W. *Org. Magn. Reson.*, **9**, 69 (1977).
73. Äyräs, P. *Org. Magn. Reson.*, **9**, 663, (1977).
74. Claret, P. A.; Osborne, A. G. *Org. Magn. Reson.*, **8**, 147 (1976).
75. Takeuchi, Y. *Org. Magn. Reson.*, **7**, 181 (1975).
76. Braun, S.; Kinkeldei, J. *Tetrahedron*, **33**, 3127 (1977).
77. Äyräs, P. *Acta. Chem. Scand. Ser. B*, **31**, 325 (1977).
78. Crecely, K. M.; Crecely, R. W.; Goldstein, J. H. *J. Mol. Spectrosc.*, **37**, 252 (1971).
79. Hayamizu, H.; Yamamoto, O. *Org. Magn. Reson.*, **13**, 460 (1980).

3

ALIPHATIC VICINAL CARBON–CARBON COUPLINGS

Aliphatic Carbon–Carbon Couplings: General

A brief examination of simple butane derivatives sets the stage for aliphatic carbon–carbon couplings. Table 3-1 lists the two-bonded ($^2J_{CC}$) and three-bonded ($^3J_{CC}$) carbon–carbon couplings for various butane compounds of the general structure $CH_3CH_2CH_2*C$.[1] In this table, analogous values of carbon–proton couplings are given in order to observe the differences between C—H and C—C couplings. These analogous carbon–proton couplings are

$$\begin{matrix} H & H \\ | & | \\ \end{matrix}$$

taken from compounds of the general structure C—C—*C whose functionality is the same at the terminal carbon as is the functionality of the C—C compound $CH_3CH_2CH_2*C$. Because the aliphatic compounds of Table 3-1 are conformationally mobile, the C—C and C—H couplings result from time-averaged conformations, and the geometries of the compared C—C and C—H couplings are not identical. Nevertheless, observations made from Table 3-1 allow generalizations that hold through a wide range of systems.

The most striking feature of the data of Table 3-1 is that three-bonded carbon–carbon couplings ($^3J_{CC}$) are *much* larger than two-bonded carbon–carbon couplings ($^2J_{CC}$). For carbon–proton couplings, three-bonded couplings ($|^3J_{CH}|$) and two-bonded couplings ($|^2J_{CH}|$) are similar. It is to be recalled that for proton–proton couplings, two-bonded couplings ($|^2J_{HH}|$) are larger than three-bonded couplings ($|^3J_{HH}|$). (In all of these comparisons, absolute values of J are used). Thus, a general trend is observed wherein ^2J becomes closer to zero as one progresses from H—H through C—H to C—C couplings. Because aliphatic $^2J_{CC}$ values are generally much smaller than aliphatic $^3J_{CC}$, $^2J_{CH}$, $^3J_{CH}$, $^2J_{HH}$, and $^3J_{HH}$ values, geminal aliphatic carbon–carbon couplings comprise a unique class with special considerations. Hence, a discussion of $^2J_{CC}$ is deferred to Chapter 5 (Special Topics) and is included with aliphatic four-bonded carbon–carbon couplings.

TABLE 3-1. LONG-RANGE CARBON–CARBON COUPLINGS OF ALIPHATIC DERIVATIVES COMPARED WITH ANALOGOUS CARBON–PROTON COUPLINGS[a]

Compound	$^2J_{CC}$	$^3J_{CC}$	$^2J_{CH}$	$^3J_{CH}$
$CH_3CH_2CH_2\overset{*}{C}OOH$ 19	1.73	3.62	6.4	5.5
$CH_3CH_2CH_2\overset{*}{C}H_2OH$ 20	<1	4.6	~4	6.4
$CH_3CH_2CH_2\overset{*}{C}H_2Cl$ 21	<1	4.8	3.7 4.2 3.9	5.7 5.63
$CH_3CH_2CH_2\overset{*}{C}H_2Br$ 22	<1	5.2	4.0	5.84
$CH_3CH_2CH_2\overset{*}{C}H_2I$ 23	<1	4.9	−5.0	5.99
$CH_3CH_2CH_2\overset{*}{C}H_2N(CH_2CH_3)_2$ 24	~0	1.9(N) 4.8		
$(CH_3CH_2CH_2)_3\overset{*}{C}OH$ 25	~0	5.0	3.8	4.5
$(CH_3CH_2CH_2)_3\overset{*}{C}H$ 26	0.8	3.8	−4.5	4.65 ~4.8
$CH_3(CH_2)_8\overset{*}{C}H_3$ 27	~0	4.0	−4.5	4.65 ~4.8

[a]Reprinted with permission from ref 1. Copyright 1982 American Chemical Society.

Further inspection of Table 3-1 shows that substituting a proton by a carbon produces a new coupling ca. 0.6–0.8 the former value in three-bonded couplings. Therefore, one sees that $^3J_{CC} = 0.7 \, ^3J_{CH}$, just as $^3J_{CH} = 0.7 \, ^3J_{HH}$, and a consistent decrease in ^3J occurs through the series H—H, C—H, and C—C vicinal couplings. The decrease in ^3J does not depend strictly upon the change of the magnetogyric ratio (which would suggest $^3J_{HH} = 0.25 \, ^3J_{CH}$ and $^3J_{CH} = 0.25 \, ^3J_{CC}$). Instead, this decrease in ^3J depends upon other factors, not understood but at least consistent. This consistent behavior suggests that a similar mechanism is operating for $^3J_{CC}$, $^3J_{CH}$, and $^3J_{HH}$, and that proton–proton or carbon–proton couplings may be used as models for carbon–carbon couplings.

Aliphatic Carbon–Carbon Couplings: The Dihedral Angular Relationship of $^3J_{CC}$

Without a doubt the greatest contribution of proton NMR to conformational analysis has been made via the application of the Karplus relationship.[2] In this relationship, $^3J_{HH}$ is maximum when the dihedral angle of the H—C—C—H linkage is 0° or 180° (8–14 Hz) and is minimum when the dihedral angle is 90° (0 Hz). Two other factors that modify the $^3J_{HH}$ value are also important. The first is the presence of a substituent X (for H—$\overset{\displaystyle X}{\overset{|}{C}}$—C—H): an electronegative substituent decreases the coupling and an electropositive substituent increases the coupling.[2] The second factor is the flaring of the hydrogens; e.g., in cyclopropene (1) the coupling is smaller than in cyclopentene (2), because the hydrogens are further apart in 1.

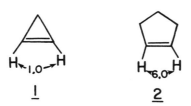

$\underline{1}$ $\underline{2}$

In carbon–carbon couplings an additional complication exists: the terminal carbon of the C—C—C—C linkage may bear substituents. As seen in Table 3-1, a terminal electronegative substituent X (on X—C—C—C—C) increases the vicinal coupling. Therefore, one would not expect the $^3J_{CC}$ values of, say, $CH_3CH_2CH_2CH_3$ and $CH_3CH_2CH_2CH_2OH$ to be equal even if the dihedral angle of the C—C—C—C linkages were the same. Accordingly, it would be desirable to develop a family of plots ($^3J_{CC}$ vs. dihedral angle) for different substituents, e.g., one plot for C—C—C—CO_2H, another for C—C—C—C—OH, etc.

Another complication anticipated in an attempt to establish a Karplus-like relationship for carbon–carbon couplings would be the orientation of a terminal substituent. That is, the question is raised: how much variation occurs in $^3J_{CC}$ as the orientation of the terminal substituent X is changed with respect to the C—C—C—C linkage? For a cis coupling (3), how much does $^3J_{CC}$ change as the orientation of the substituent changes from cisoid (3a) to transoid (3b)? For a trans coupling (4), moreover, the same question pertains to the conformations 4a and

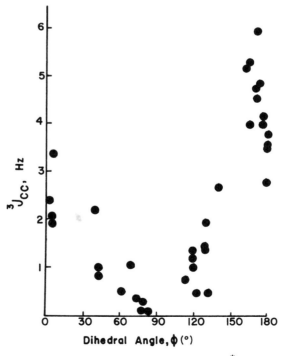

4b. One anticipates a greater difference in $^3J_{CC}$ for **3** than **4**, because of potential through-space effects in **3**, but in either case one wonders whether the orientation effect of the substituent is so great as to eclipse the overall trends of the dihedral angle relationship and to limit their applicability.

Finally, another difficulty in $^3J_{CC}$ studies not observed in $^3J_{HH}$ work is that more through-space interactions are possible in carbon–carbon couplings. These interactions arise from the additional carbon and hydrogen orbitals on the terminal atoms of the C—C—C—C linkage.

Figure 3-1. $^3J_{CC}$ vs. dihedral angle ϕ of C—C—C—$\overset{*}{C}O_2H$ compounds.

$^3J_{CC}$ vs. Dihedral Angle of C—C—C—C

The first of these considerations regarding $^3J_{CC}$ vs. dihedral angle is now addressed, i.e., the question of families of curves for $^3J_{CC}$. Table 3-2 lists the J_{CC} values for aliphatic compounds, from which Figures 3-1, 3-2, and 3-3 were developed. These figures portray the $^3J_{CC}$ values vs. dihedral angle for, respectively, C—C—C—*CO$_2$H, C—C—C—*C—OH, and C—C—C—*CHO for a wide variety of rigid structural types including norbornanes, adamantanes, and cycloalkanes. These plots include 81 points encompassing dihedral angles from near 0° to 180°. The values for the dihedral angles were obtained from a combination of available electron diffraction and X-ray crystallographic data of similar compounds and of force field calculations.[3] Each of the three figures exhibits a general trend similar to that for proton–proton couplings, i.e., with maxima near 0° and 180° and a minimum near 90°. For carboxylic acids (Figure 3-1) the maximum at 180° is greater than that at 0°, just as for proton–proton couplings. In contrast, for alcohols (Figure 3-2) and aldehydes (Figure 3-3) the maximum at 0° is greater. Figures 3-1, 3-2, and 3-3 attest to the validity of the Karplus-like relationship for carbon–carbon couplings for a variety of structural

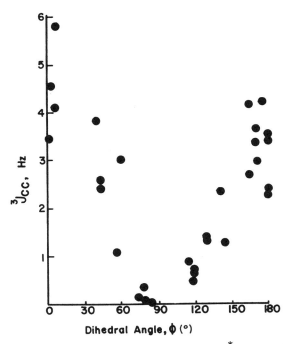

Figure 3-2. $^3J_{CC}$ vs. dihedral angle ϕ of C—C—C—*CH$_2$OH compounds.

TABLE 3-2. CARBON-CARBON COUPLING CONSTANTS (AND CARBON-13 CHEMICAL SHIFTS) OF ALIPHATIC COMPOUNDS

J_{CC}(Hz) to labeled carbon (δ_{C-13}, ppm) of carbon no.:

Compound	1	2	3	4	5	6	7	8	9	10	11	Other	Reference
5	0.44 (48.1)	55.32 (63.5)	1.86 (42.7)	0.78 (48.8)	<0.5 (25.6)	4.78 (37.8)	<0.6 (42.7)	0.98 (18.5)	(179.4)	3.35 (26.3)	1.39 (28.0)		8
6	0.39 (48.3)	56.57 (61.2)	0.24 (40.2)	1.95 (49.1)	<0.7 (26.0)	1.06 (27.4)	5.96 (45.6)	<0.7 (20.8)	(179.9)	0.98 (32.0)	2.09 (22.7)		8
7	0.38 (47.1)	38.49 (58.6)	1.10 (41.5)	0.88 (49.4)	<0.2 (25.6)	2.97 (39.3)	<0.3 (43.0)	2.57 (19.0)	(62.4)	5.76 (24.1)	0.68 (28.5)		8
8	0.55 (47.5)	39.97 (57.9)	<0.2 (38.8)	1.32 (49.0)	<0.2 (26.2)	2.99 (28.1)	3.64 (45.3)	0.18 (21.3)	(60.7)	0.61 (38.7)	4.07 (21.0)		8

Structure												
9 (CHO)	<0.2 (48.3)	41.12 (68.9)	1.58 (43.5)	0.61 (48.4)	<0.2 (25.6)	3.42 (37.6)	<0.4 (43.4)	1.98 (18.1)	(205.5)	5.37 (25.7)	1.10 (28.3)	∞
10 (CHO)	0.76 (48.9)	42.19 (67.1)	<0.2 (41.4)	1.28 (49.3)	<0.2 (26.1)	2.69 (28.7)	4.33 (45.4)	<0.2 (20.7)	(206.6)	0.92 (32.6)	3.85 (22.8)	∞
11 (CO₂H)	0.92 (50.1)	56.02 (51.7)	1.44 (33.1)	1.41 (46.1)	0.42 (28.2)	4.54 (40.2)	<0.18 (48.4)	0.84 (15.3)	(176.7)	<0.11 (19.9)	<0.17 (20.6)	∞
12 (CO₂H)	1.25 (49.9)	57.67 (50.2)	<0.2 (32.0)	1.43 (46.0)	<0.2 (28.4)	2.17 (31.1)	4.25 (50.3)	<0.26 (14.6)	(176.9)	<0.11 (19.1)	0.31 (19.5)	∞
13 (CH₂OH)	0.1 (47.7)	38.57 (51.3)	0.70 (35.1)	1.39 (46.0)	0.33 (28.0)	3.37 (40.6)	0.1 (47.4)	2.38 (13.0)	(66.4)	0.73 (20.9)	<0.11 (20.9)	∞

TABLE 3-2. (continued)

J_{CC}(Hz) to labeled carbon (δ_{C-13}, ppm) of carbon no.:

Compound	1	2	3	4	5	6	7	8	9	10	11	Other	Reference
14	<0.10 (47.6)	40.33 (46.6)	0.81 (34.9)	1.26 (46.0)	<0.15 (27.4)	3.81 (29.5)	2.69 (49.7)	<0.18 (15.3)	(65.1)	<0.11 (18.9)	0.33 (19.3)		8
15	1.81 (50.0)	74.3 (168.9)	1.41 (42.9)	2.12 (47.5)	≤0.7 (25.6)	2.09 (35.7)	2.44 (44.4)	2.36 (18.4)	(97.4)	2.59 (29.7)	1.76 (26.2)		8
16	1.39 (51.5)	74.30 (159.5)	1.43 (37.1)	2.82 (45.0)	0.29 (28.1)	2.31 (35.3)	2.16 (47.3)	2.38 (12.5)	(101.3)	≤0.15 (19.7)	0.18 (19.0)		8
17	57.6 (40.9)	1.42 (39.7)	3.23 (29.1)	0.51 (37.5)									8

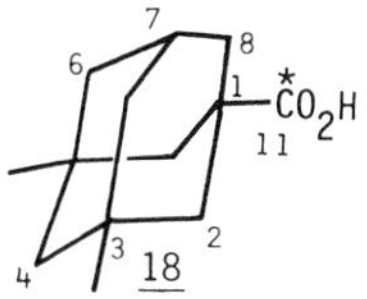

18 (CO₂H)

57.36 (42.6) 1.40 (45.5) 3.37 (31.2) 0.54 (51.2) 0.45 (43.3) 3.51 (29.8) 1.38 (38.0) (180.2) C-12 0.29 (30.7) 8

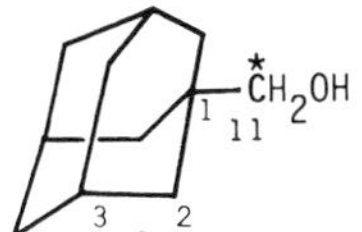

19 (CH₂OH)

40.3 (34.7) <0.17 (39.4) 3.31 (28.4) 0.55 (37.5) (73.2) 10

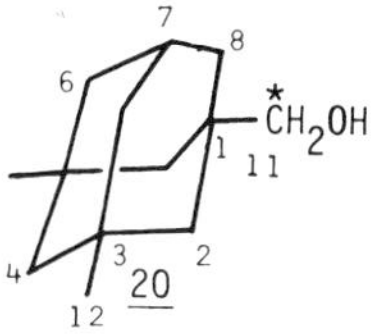

20 (CH₂OH)

39.71 (36.9) <0.08 (46.1) 3.39 (31.3) 0.46 (52.0) 0.45 (44.0) 3.52 (30.1) <0.09 (38.4) (72.8) C-12 <0.18 (31.0) 8

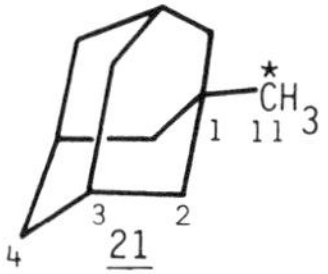

21 (CH₃)

36.97 (29.7) <0.2 (44.7) 3.26 (28.9) 0.48 (37.0) <0.3 (27.8) (31.4) 8

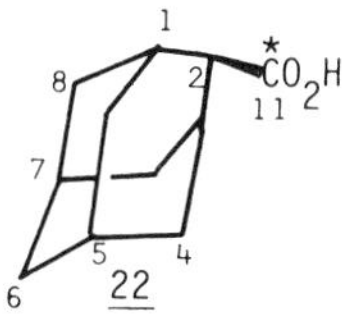

22 (CO₂H)

1.09 (29.8) 56.98 (49.8) <0.2 (37.6) 0.5 (27.8) 4.44 (38.2) 8

TABLE 3-2. (continued)

| Compound | J_{CC}(Hz) to labeled carbon (δ_{C-13}, ppm) of carbon no.: | | | | | | | | | | | | Reference |
	1	2	3	4	5	6	7	8	9	10	11	Other	
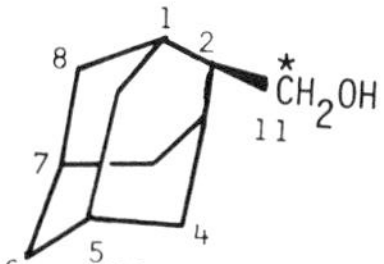 *CH$_2$OH adamantyl, 23	0.27 (29.4)	38.3 (47.0)		0.85 (32.1)	0.13 (28.2)	<0.13 (38.4)	0.35 (28.6)	3.59 (39.1)			(64.7)		11
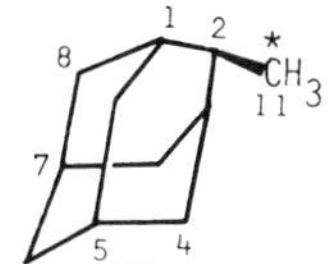 *CH$_3$ adamantyl, 24	0.60 (33.9)	35.8 (39.2)		0.91 (31.4)	0.19 (28.6)	0.17 (38.7)	0.35 (28.3)	3.42 (39.6)			(18.9)		8
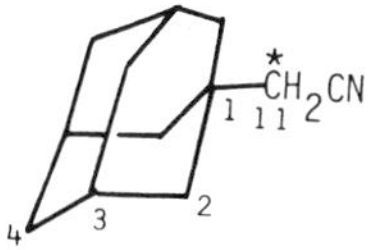 *CH$_2$CN adamantyl, 25		<0.07 (41.9)	3.82 (28.4)	0.53 (36.4)							(32.3)		12
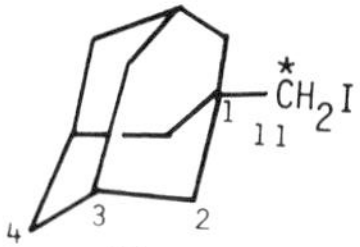 *CH$_2$I adamantyl, 26	36.31 (32.4)	<0.2 (42.1)	3.68 (28.7)	0.55 (36.6)							(26.7)		12

74

27 CH$_2$Cl	3.4						12
28 CO$_2$H	56.5 (44.5)	0.6 (30.5)	2.7 (26.4)				9
29 CO$_2$H	55.77 (43.2)	1.50 (29.4)	4.10 (25.9)	0.50 (26.4)	(180.4)		8
30 CO$_2$H	~0.0 (39.8)	~0.0 (28.6)		5.6 (30.6)	59.2 (54.2)		9
31 CO$_2$H	1.0 (41.7)	56.1 (47.1)	1.8 (34.3)	~0.5 (36.9)	~0.6 (29.2)	4.9 (30.0)	~0.0 (37.0)	9
32 CO$_2$H						2.4 (15.8)	9

TABLE 3-2. (continued)

J_{CC}(Hz) to labeled carbon (δ_{C-13}, ppm) of carbon no.:

Compound	1	2	3	4	5	6	7	8	9	10	11	Other	Reference
33	~0.4 (42.1)	58.5 (48.0)	~0.0 (35.9)	~0.5 (44.0)	~0.3 (23.2)	~0.5 (24.5)	5.2 (41.1)		1.9 (14.9)				9
34	38.8 (42.5)	<0.5 (29.6)	2.37 (25.9)			(66.7)							10
35	38.3 (41.0)	<0.5 (30.3)	4.2 (26.5)	<0.5 (27.3)			(68.1)						10
36	<0.5 (37.7)	1.04 (27.8)			4.15 (31.0)		40.9 (57.0)	(61.1)					10
37	<0.5 (41.6)	<0.5 (135.6)	<0.5 (135.6)	<0.5 (49.6)	<0.5 (35.8)	<0.5 (46.1)	4.5 (44.9)	(62.7)	4.57 (14.1)				10

76

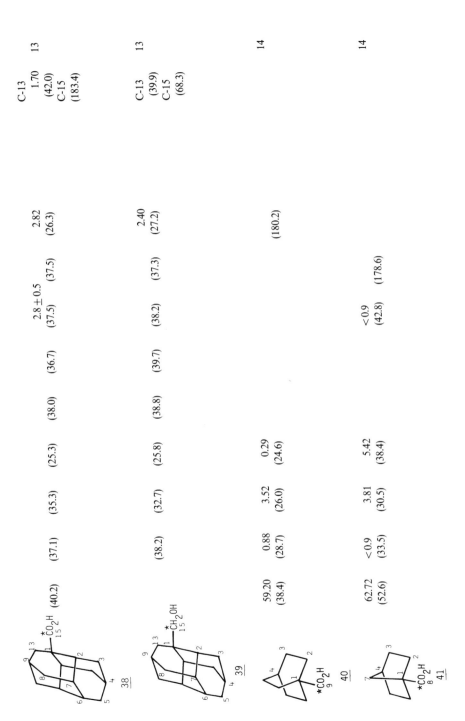

38

(37.1) (35.3) (25.3) (38.0) (36.7) 2.8±0.5 (37.5) 2.82 (26.3)

C-13
1.70 (42.0)
C-15 (183.4)   13

39

(38.2) (32.7) (25.8) (38.8) (39.7) (38.2) (37.3) 2.40 (27.2)

C-13 (39.9)
C-15 (68.3)   13

40

59.20 (38.4) 0.88 (28.7) 3.52 (26.0) 0.29 (24.6) (180.2)   14

41

62.72 (52.6) <0.9 (33.5) 3.81 (30.5) 5.42 (38.4) <0.9 (42.8) (178.6)   14

TABLE 3-2. (continued)

J_{CC}(Hz) to labeled carbon (δ_{C-13}, ppm) of carbon no.:

Compound	1	2	3	4	5	6	7	8	9	10	11	Other	Reference
(42)	65.41 (52.7)	1.87 (29.6)	3.39 (28.1)	8.49 (37.1)	0.52 (42.2)		(176.2)						14
(43)	40.40 (32.3)	0.58 (27.6)	3.39 (25.6)	0.22 (24.5)					(71.8)				14
(44)	42.22 (50.1)	1.71 (31.8)	3.51 (30.5)	4.15 (37.3)			1.02 (40.6)	(66.9)					14
(45)	42.69 (53.5)	3.03 (28.2)	3.09 (28.2)	7.81 (36.7)	<0.12 (39.6)		(65.2)						14

78

Structure								
46 *CN (4)	78.59 (10.2)	1.95 (20.6)	2.35 (22.1)	(120.9)	2.68 (21.9)	0.99 (24.8)		∞
47 *CO$_2$H (4)	73.71 (27.0)	1.52 (23.8)	1.48 (22.6)	(177.8)	1.57 (18.8)	1.22 (27.0)		∞
48 *C–NH$_2$, O (4)	63.4 (28.9)	1.5 (22.5)	1.4 (20.5)	(178.2)	1.5 (19.0)	1.0 (27.1)		∞
49 *CHO (4)	53.70 (35.5)	1.27 (25.1)	1.32 (22.3)	(200.3)	2.65 (18.8)	0.74 (25.9)		∞
50 *CH$_2$OH (4)	48.40 (27.2)	1.19 (16.0)	0.61 (18.3)	(63.1)	3.40 (19.8)	0.46 (27.6)		∞
51 *C–CH$_3$, O (4,7)	52.54 (35.5)	1.66 (26.2)	1.89 (23.0)	(205.5)	1.42 (18.1)	0.98 (27.2)	41.55 (32.0)	∞
52 H *C CH$_3$ OH (4,7)	48.71 (33.0)	1.69 (16.2)	<0.2 (18.3)	(70.1)	3.23 (20.0)	0.37 (27.4)	39.17 (23.2)	∞

79

TABLE 3-2. (continued)

J_{CC}(Hz) to labeled carbon (δ_{C-13}, ppm) of carbon no.:

Compound	1	2	3	4	5	6	7	8	9	10	11	Other	Reference
52a	1.10	3.45	0.34	39.17	<0.2	<0.1							8
53	47.69 (32.9)	<0.4 (16.2)	1.83 (18.8)	(69.8)	2.72 (20.3)	0.73 (27.5)	38.88 (23.9)						8
53a			3.23	38.88	<0.2	<0.1							8
54	47.82 (35.6)	1.44 (16.3)	1.71 (15.3)	(70.1)	1.04 (19.7)	0.98 (20.0)	39.83 (29.5)	39.22 (32.2)					8
54a	3.11	2.62	1.03	39.2	<0.2	<0.6							8

Compound											Ref
55	40.03 (83.6)	1.76 (46.1)	1.46 (55.1)	3.14 (34.1)	2.53 (54.5)	(22.8)	1.71 (26.8)	0.49 (22.6)	<0.2 (33.0)	<0.2 (33.7)	8
56	43.95 (144.5)	2.93 (134.1)	5.01 (42.2)	1.34 (55.3)	3.17 (46.5)	(12.0)	<0.5 (28.9)		0.61 (30.8)		8
57	54.17 (51.8)	1.01 (45.7)	0.81 (52.6)	1.04 (37.6)	<0.28 (49.4)	(183.9)	<0.28 (25.2)	1.99 (14.7)	<0.28 (30.4)	<0.28 (31.7)	8
58	1.37 (51.9)	40.82 (80.0)	0.64 (43.0)	0.63 (49.0)	≤0.2 (25.7)	0.97 (29.5)	≤0.2 (40.9)	2.55 (16.9)	(23.1)	5.32 (27.2)	≤0.2 (22.1)	8
59	2.7	41.1	1.5	0.9	<0.35	1.8	<0.35				15

Structures:

55 — OH, atoms numbered 1, 2, 3, 4, 5, 6, 7, 8, 9, 10, with *

56 — atoms numbered 1, 2, 3, 4, 5, 6, 7, 8, 9, with *

57 — CO_2H, atoms numbered 1, 2, 3, 4, 5, 6, 7, 8, 9, 10, with *

58 — OH, atoms numbered 1, 2, 3, 4, 5, 6, 7, 8, 9, 10, 11, with *

59 — CH_3, OH, atoms numbered 1, 2, 3, 4, 6, 7, with *

81

TABLE 3-2. (continued)

J_{CC}(Hz) to labeled carbon (δ_{C-13}, ppm) of carbon no.:

Compound	1	2	3	4	5	6	7	8	9	10	11	Other	Reference
60	41.6	2.25	3.2										15
61	0.6 (46.4)	52.3 (45.5)	<0.6 (42.1)	<0.6 (53.3)	<1.0 (30.9)	3.1 (78.8)	4.9 (33.8)		2.2 (19.5)				8
62	1.2 (47.9)	52.3 (41.8)	<0.6 (46.4)	<0.6 (52.6)	<1.0 (26.9)	3.1 (79.8)	3.7 (37.4)		3.1 (20.2)				8
63	0.7 (46.5)	34.7 (46.3)	≤0.3 (43.8)	0.7 (50.4)	≤0.2 (37.7)	0.4 (88.6)	2.7 (34.3)	(73.5)	2.6 (20.8)				8

64

0.5 (44.6) 34.9 (47.0) ≤0.2 (42.5) 0.7 (47.6) ≤0.15 (42.3) 0.6 (76.8) 2.7 (33.4) (73.1) 2.8 (20.5) 8

90

35.9 (47.0) (218.2) 1.1 (39.3) 2.0 (27.5) <0.5 (36.4) 16

91

34.5 (34.6) (74.5) 0.7 (36.6) 1.3 (27.1) <0.5 (37.6) 1.7 (27.6) 1.8 (31.0) 16

92

33.9 (35.1) (59.5) <0.5 (37.8) 1.3 (27.4) <0.5 (38.0) 1.6 (27.8) 1.5 (30.8) 16

TABLE 3-2. (continued)

J_{CC}(Hz) to labeled carbon (δ_{C-13}, ppm) of carbon no.:

Compound	1	2	3	4	5	6	7	8	9	10	11	Other	Reference
93	31.0 (31.3)	(57.0)		<0.5 (37.3)	1.5 (27.4)	<0.5 (37.6)	1.6 (27.6)	1.3 (30.7)					16
94	33.3 (32.0)	(53.0)		<0.5 (37.2)	1.4^a (27.2)	<0.5 (37.6)	<0.5^a (27.1)	1.5 (32.3)					16
95	32.1 (29.5)	(49.5)		<0.5 (38.1)	1.4^a (27.5)	<0.5 (37.4)	<0.5^a (27.5)	1.2 (33.6)					16
96	36.5 (50.9)	(212.9)	35.5 (43.3)	1.2 (27.9)	1.4 (38.0)	1.1 (54.4)	2.0 (17.3)	2.4 (24.2)					17

84

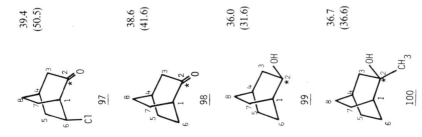

97

39.4 (50.5)   (211.4)   35.6 (43.9)   1.6 (28.4)   1.8 (37.7)   2.7 (55.2)   1.8 (22.7)   2.7 (23.3)   17

98

38.6 (41.6)   (217.3)   34.6 (44.6)   1.5 (27.9)   2.2 (24.7)   1.7 (23.3)   17

99

36.0 (31.6)   (69.5)   34.7 (37.5)   1.0 (24.9)   2.4 (25.7)   1.9 (18.7)   0.7 (23.8)   2.4 (24.6)   17

100

36.7 (36.6)   (71.7)   35.4 (44.1)   1.2 (26.2)   1.7 (24.4)   1.7 (21.1)   (23.3)   1.8 (24.6)   17

TABLE 3-2. (continued)

J_{CC}(Hz) to labeled carbon (δ_{C-13}, ppm) of carbon no.:

Compound	1	2	3	4	5	6	7	8	9	10	11	Other	Reference
101	2.2 (32.0)	(26.0)	(28.3)	(58.5)	35.0 (53.5)	(216.2)	36.7 (42.9)	3.9 (31.5)					17
102	2.3 (32.0)	(30.5)	0.7 (18.7)	2.0 (30.3)	37.4 (45.9)	(221.2)	35.6 (43.3)	4.4 (37.0)					17
103	1.7 (34.3)	(32.6)	0.4 (19.5)	2.2 (26.8)	35.6 (39.3)	(75.0)	36.3 (38.0)	6.1 (37.9)					17
104	(30.5)	3.4 (31.3)	0.4 (125.7)	(126.9)	(24.7)	2.6 (28.4)	55.1 (40.9)	(129.8)					17

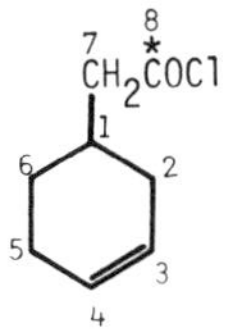 105	1.9 (30.5)	3.9 (30.2)	0.3 (124.4)	(126.4)	(24.0)	4.0 (27.4)	52.9 (52.8)	(172.0)		17
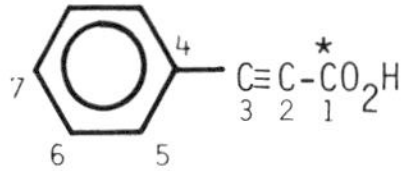 106	(156.6)	119.3 (81.8)	18.4 (86.7)	2.0 (120.8)	1.0 (130.0)	(133.7)	(131.7)			18
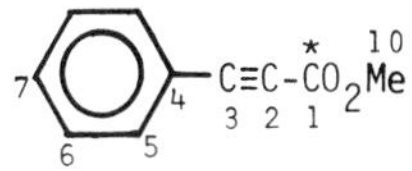 107	(153.2)	127.7 (81.8)	19.5 (87.8)	2.4 (120.9)	1.2 (129.9)	(134.4)	(132.1)		1.8 (54.1)	18
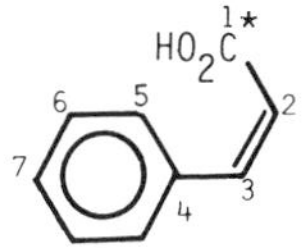 108	(169.0)	72.0 (121.1)	(143.1)	2.0 (136.4)	(130.6)	(129.0)	(129.8)			18
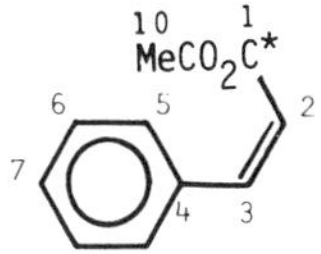 109	(166.4)	75.7 (119.2)	(143.2)	2.9 (134.7)	(129.6)	(127.9)	(128.9)		2.2 (51.2)	18

TABLE 3-2. (continued)

J_{CC}(Hz) to labeled carbon (δ_{C-13}, ppm) of carbon no.:

Compound	1	2	3	4	5	6	7	8	9	10	11	Other	Reference
110	(170.4)	72.6 (119.3)	1.4 (146.4)	6.6 (135.7)	(129.1)	(129.9)	(131.3)						18
111	(168.5)	76.5 (119.1)	2.2 (146.0)	7.4 (135.6)	(129.3)	(130.1)	(131.5)			2.2 (52.8)			18
112	(176.7)	55.2 (36.6)	1.8 (31.8)	3.5 (141.6)	(129.4)	(129.2)	(127.1)						18
113	(172.7)	57.4 (35.2)	1.3 (30.5)	3.6 (140.2)	(128.1)	(127.9)	(125.9)			2.2 (51.0)			18
114	(157.5)	123.0 (72.6)	19.3 (91.9)	1.5 (18.4)	(28.3)	(27.4)	(31.2)	(22.4)	(13.6)				18

115: CO_2Me (154.1), 126.2 (72.7), 19.4 (89.7), (18.4), (28.3), (27.4), (31.0), (22.3), (13.8), 1.8 (52.3), 18

116: CO_2H (169.8), 70.8 (120.9), (151.5), (30.1), (30.1), (29.9), (32.9), (23.6), (14.4), 18

117: CO_2Me (166.8), 73.6 (119.0), (150.9), (28.9), (28.9), (28.9), (31.6), (22.5), (13.9), 2.7 (50.8), 18

118: CO_2H (170.1), 71.6 (122.5), 0.7 (151.2), 6.4 (33.1), (29.9), (29.2), (32.7), (23.6), (14.4), 18

119: CO_2Me (166.8), 74.2 (120.7), 0.9 (148.3), 7.2 (32.0), (28.6), (27.8), (31.4), (22.3), (13.7), 2.7 (51.0), 18

120: CO_2H (177.8), 55.2 (34.9), 1.7 (25.9), 2.9 (30.2), 1.0 (30.3), (30.2), (32.9), (23.6), (14.5), 18

121: CO_2Me (173.7), 57.6 (33.7), 1.8 (24.6), 2.5 (28.9), (28.9), (28.9), (31.5), (22.3), (13.6), 2.9 (50.8), 18

TABLE 3-2. (continued)

Compound	J_{CC}(Hz) to labeled carbon ($\delta_{C\text{-}13}$, ppm) of carbon no.:											Other	Reference
	1	2	3	4	5	6	7	8	9	10	11		
122	(167.6)		7.8 (135.5)										18
123	(164.3)		7.7 (139.9)										18
124	(168.4)		5.5 (136.6)										18
125	(168.4)		3.4 (125.9)										18

90

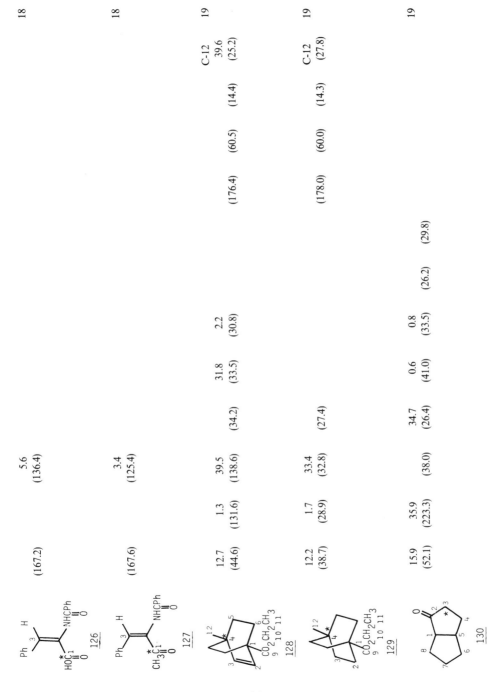

Compound								Ref
126	(167.2)		5.6 (136.4)					18
127	(167.6)		3.4 (125.4)					18
128	12.7 (44.6)	1.3 (131.6)	39.5 (138.6)	(34.2)	31.8 (33.5)	2.2 (30.8)	(176.4) (60.5) (14.4) C-12 39.6 (25.2)	19
129	12.2 (38.7)	1.7 (28.9)	33.4 (32.8)	(27.4)			(178.0) (60.0) (14.3) C-12 (27.8)	19
130	15.9 (52.1)	35.9 (223.3)	(38.0)	34.7 (26.4)	0.6 (41.0)	0.8 (33.5) (26.2) (29.8)		19

91

TABLE **3-2.** (continued)

Compound	for header

J_{CC}(Hz) to labeled carbon (δ_{C-13}, ppm) of carbon no.:

Compound	1	2.	3	4	5	6	7	8	9	10	11	Other	Reference
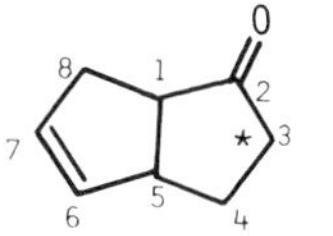 131	15.5 (49.3)	36.8 (223.3)	(36.3)	35.2 (25.3)	0.8 (47.6)	(133.4)	(130.8)	(37.3)					19
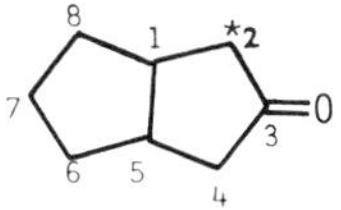 132	36.0 (39.6)	(44.7)	36.6 (220.6)			(33.5)	1.8 (25.6)						19
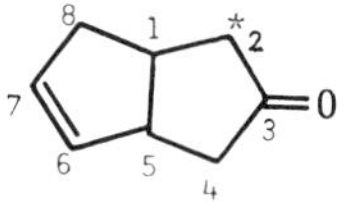 133	34.7 (37.1)	(44.8)	36.8 (219.5)	17.0 (42.6)	(46.2)	(133.8)	(130.1)	(40.1)					19
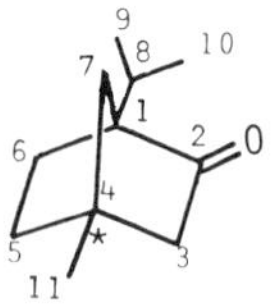 134	2.4 (63.4)	1.0 (218.1)	33.0 (52.8)	(41.1)	31.9 (35.1)	(28.2)	30.8 (44.7)	3.4 (26.5)	(19.3)	(18.4)	39.7 (21.1)		19

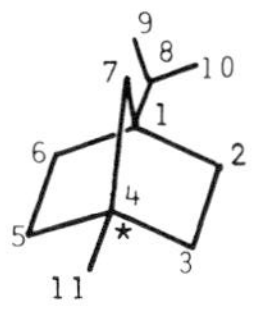 6.6 (52.5)	(32.7)	32.4 (37.6)	(43.7)	32.4 (37.6)	(32.7)	32.3 (48.3)	3.9 (32.6)	(19.1)	(19.1)	39.2 (21.4)
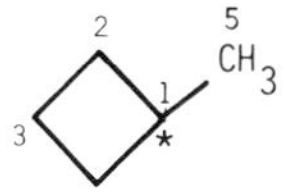 (31.6)	29.1 (30.4)	8.1 (18.6)		36.1 (22.1)						
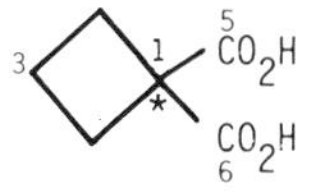 (53.0)	27.3 (29.4)	7.3 (16.6)		58.7 (173.9)						
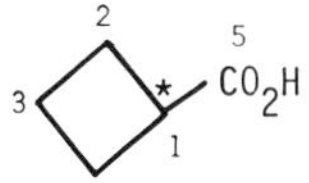 (38.5)	27.5 (25.5)	8.3 (18.8)		58.1 (182.5)						
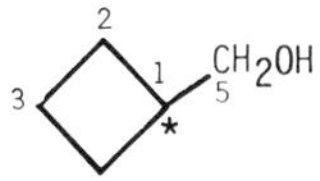 (37.6)	28.8 (24.8)	8.1 (18.6)		39.3 (67.0)						

135

136

137

138

139

19

20

20

20

20

TABLE 3-2. (continued)

J_{CC}(Hz) to labeled carbon ($\delta_{C\text{-}13}$, ppm) of carbon no.:

Compound	1	2	3	4	5	6	7	8	9	10	11	Other	Reference
140	(34.2)	28.4 (24.4)	8.3 (18.3)		39.6 (74.1)								20
141		29.52	9.66										21
142			9.0										22
143			15.2										22
144			+16.1										23

94

$\overset{*}{\underset{1}{C}H_3}O\underset{3}{C}H_3$ **145**   −2.4   23

$\overset{*}{\underset{1}{C}H_3}-Hg-\underset{3}{C}H_3$ **146**   +22.4   23

$\underset{2}{C}H_3O\overset{\overset{O}{\|}}{\underset{1}{C}}Cl$ **147**   −2.8   24

178a   37.00   0.14^b 3.38   0.45^b 25

178b   32.23   0.89   25

TABLE 3-2. (continued)

Compound	J_{CC}(Hz) to labeled carbon (δ_{C-13}, ppm) of carbon no.:											Other	Reference
	1	2	3	4	5	6	7	8	9	10	11		
179a	35.2			2.6	1.7								26
179b				33.0									26
180a	35.40 (26.6)	(31.7)	3.42 (29.2)	(33.4)			(23.3)	(17.4)					27
180b	34.64 (26.6)	(31.7)	<1 (29.2)	(33.4)			(17.4)	(23.3)					27

96

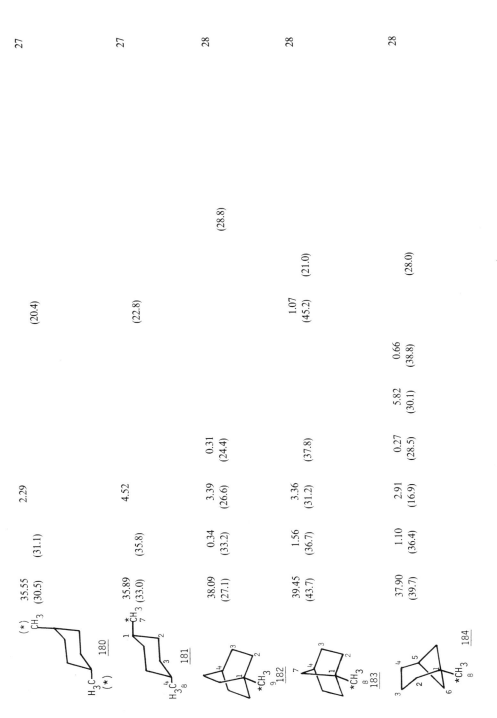

TABLE 3-2. (continued)

| Compound | J_{CC}(Hz) to labeled carbon (δ_{C-13}, ppm) of carbon no.: | | | | | | | | | | | | |
	1	2	3	4	5	6	7	8	9	10	11	Other	Reference
 185	39.65 (48.5)	2.44 (33.0)	2.77 (29.2)	7.50 (36.8)	<0.2 (43.8)		(19.8)						28
 186	37.62 (42.2)	<0.2 (52.2)	9.71 (27.4)	(52.2)		(19.1)							28
 187	(79.7)	37.1 (41.2)	1.22 (24.2)			39.8 (28.2)							29

98

188

1.39 (51.8) 40.70 (76.7) 2.54 (48.5) 1.36 (44.6) <0.6 (33.6) 3.08 (33.5) 1.75 (48.4) <0.4 (11.6) (27.9) <0.1 (19.7) 0.51 (18.8) 8

[a]Original assignments reversed on the basis of expected negative through-space contributions.
[b]Original assignments reversed on the basis of compound **21**.

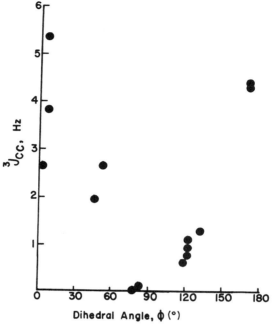

Figure 3-3. $^3J_{CC}$ vs. dihedral angle ϕ of C—C—C—$\overset{*}{C}$HO compounds.

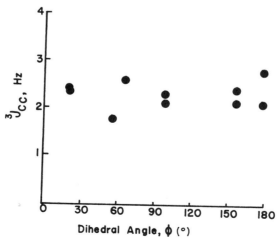

Figure 3-4. $^3J_{CC}$ vs. dihedral angle of ϕ of C—C—C=$\overset{*}{C}$H$_2$ compounds.

types and suggest that $^3J_{CC}$ can be reliably used in conformational analysis. At $0°$ and $180°$ there is considerable spread in the values of $^3J_{CC}$, however, and this matter is considered below.

For carbon fragments including a terminal olefin (i.e., for C=C—C—C) the Karplus-like relationship is not obeyed. Vicinal couplings of this type are seen in compounds **15** and **16** of Table 3-2. From these two compounds Figure 3-4 was developed, showing $^3J_{CC}$ vs. dihedral angle for 10 points. The result is a plot where the $^3J_{CC}$ value is not related to the dihedral angle. The ostensible reason is that at dihedral angles near $90°$, hyperconjugative interaction (as shown in **61**) compensates for the decrease of the "normal" contributions to $^3J_{CC}$ as depicted in Figures 3-1, 3-2, and 3-3. Theoretical calculations have attested to this hyperconjugative effect in carbon–carbon couplings,[4] and analogous pheno-

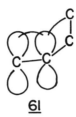

61

mena have been observed in carbon–proton[5,6] and even in proton–proton

couplings (for the fragment =C—C[7]).

$^3J_{CC}$ vs. Orientation of a Terminal Substituent

The question of the effect of the orientation of a terminal substituent is now considered. In contrast to the relative ease of investigating the dependence of $^3J_{CC}$ upon the dihedral angle of the rigid C—C—C—C framework, it is more difficult to acquire empirical data relating $^3J_{CC}$ to the orientation of a terminal substituent of the fragment C—C—C—C—X, because the substituent is generally free to rotate. To address this question, a preliminary theoretical approach is therefore helpful. Tables 3-3, 3-4, and 3-5 show calculations[8] for the three classes of compounds, respectively: alcohols (C—C—C—C—OH), aldehydes (C—C—C—CHO) and carboxylic acids (C—C—C—CO_2H). For each of these three classes of compounds, the calculated dependence of $^3J_{CC}$ is given upon the dihedral angle of the four atom fragment including the substituent, e.g., C—C—C—X. This dependence is given in two general geometries of the carbon framework (C—C—C—C), both when the vicinal coupling nuclei are mutually close (dihedral angle $= 0°$) and when they are not (dihedral angle $= 120°$). To demonstrate the effect of through-space interactions,

TABLE 3-3. CALCULATED (INDO-FPT)[a] VICINAL CARBON–CARBON COUPLING CONSTANTS FOR 3-METHYL-1-BUTANOL[8]

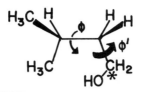

ϕ' (degrees)[b]	$^3J_{CC}$(Hz) for $\phi = 0°$ [c]		$^3J_{CC}$(Hz) for $\phi = 120°$ [d]	
	Normal	Cutoffs[e]	Normal	Cutoffs[e]
0	0.6	1.5	1.8	1.6
45	2.3	1.6	1.9	1.6
90	6.1	2.2	1.6	1.7
135	7.1	2.6	1.4	2.0
180	6.4	2.3	1.5	1.8
225	7.6	2.5	1.5	2.3
270	6.6	2.6	1.6	1.8
315	2.3	1.6	1.4	1.7

[a] Intermediate neglect of differential overlap finite perturbation.
[b] Defined by the dihedral angle of C—C—C—O, viewed clockwise from the terminally substituted carbon.
[c] That is, coupling of the CH_2OH carbon to the cis methyl.
[d] That is, coupling of the CH_2OH carbon to the trans methyl.
[e] These calculations were performed with overlap integrals reduced to zero between the coupling atoms and all associated atoms, i.e., between CH_3— and —CH_2OH.

these calculations are also given with these interactions cut off[30] between the coupling vicinal atoms and associated atoms, that is, between the methyl groups and the functionalized group.

Inspection of Tables 3-3, 3-4, and 3-5 bears out a prediction that when the coupling atoms are mutually close (cis conformation, $\phi = 0°$) a strong dependence exists of $^3J_{CC}$ upon the orientation of the terminal substituent (upon ϕ'), but when the coupling atoms are far apart ($\phi = 120°$), little such dependence exists. This dependence of $^3J_{CC}$ upon ϕ' is predominately a through-space phenomenon, as attested by the calculations when these through-space interactions are removed. Thus, for example for 3-methyl-1-butanol (Table 3-3) in the cis conformation ($\phi = 0°$) the normal calculated values of $^3J_{CC}$ vary over a wide range of 0.6–7.6 Hz; when through-space interactions are removed, the range narrows to 1.5–2.6 Hz.

Tables 3-3, 3-4, and 3-5 inform us that the vicinal carbon–carbon coupling is generally small when the oxygen substituent is directed toward the coupled atom (i.e., a cisoid conformation), whether the oxygen is a hydroxyl group or a carbonyl group. When the hydroxyl or carbonyl group is directed away from the coupled carbon atom (i.e., a transoid conformation), the $^3J_{CC}$ value is large. This

TABLE 3-4. CALCULATED (INDO-FPT) VICINAL CARBON–CARBON COUPLING CONSTANTS FOR 3-METHYLBUTANAL[8]

ϕ' (degrees)[a]	$^3J_{CC}$(Hz) for $\phi=0°$ [b]		$^3J_{CC}$(Hz) for $\phi=120°$ [c]	
	Normal	Cutoffs[d]	Normal	Cutoffs[d]
0	2.5	4.9	2.7	2.0
45	2.6	5.1	2.5	2.0
90	3.1	5.5	2.5	1.9
135	6.1	6.3	2.3	2.0
180	8.4	6.9	2.1	2.3
225	6.0	6.2	2.0	2.0
270	3.1	5.5	2.1	2.2
315	2.6	5.1	2.2	2.0

[a] Defined by the dihedral angle of C—C—C=O, viewed clockwise from the terminally substituted carbon.
[b] That is, coupling of the CHO carbon to the cis methyl.
[c] That is, coupling of the CHO carbon to the trans methyl.
[d] These calculations were performed with overlap integrals reduced to zero between the coupling atoms and all associated atoms, i.e., between CH$_3$— and —CHO.

through-space effect is not only attenuating (for the cisoid conformation) but is augmenting (for the transoid conformation), owing to various combinations of orbital interactions, including protons, carbons, and heteroatoms (for a more detailed discussion of these complications, see reference 12). It is observed from Tables 3-3, 3-4, and 3-5 both the attenuating and the augmenting effect is similar for both the hydroxyl group and the carbonyl group. For example, the range of normal calculated $^3J_{CC}$ values for 3-methylbutanal (Table 3-4) is about the same as for 3-methyl-1-butanol (Table 3-3). For 3-methylbutanoic acid (Table 3-5), the geometrically opposed hydroxyl and carbonyl groups tend to offset each other's effect, and accordingly the normal calculated $^3J_{CC}$ values are restricted to the range 2.0–3.5 Hz. Because of this narrow range of $^3J_{CC}$, the carboxyl group appears to be one of the better functionalities to avoid heteroatom through-space complications and therefore to adopt for conformational analysis of carbon frameworks.

It is desirable to test the calculations of Tables 3-3, 3-4, and 3-5 with empirical data. One set of compounds that affords such data includes the lactones and cyclic ethers **61–64** (from Table 3-2). In these compounds the functionality is held roughly orthogonal to the coupling path of the C—C—C—C framework, both for cis (**62 and 64**) and for trans (**61 and 63**) vicinal couplings. Couplings

TABLE 3-5. CALCULATED (INDO-FPT) VICINAL CARBON–CARBON COUPLING CONSTANTS FOR 3-METHYLBUTANOIC ACID[8]

ϕ' (degrees)[a]	$^3J_{CC}$(Hz) for $\phi = 0°$ [b]		$^3J_{CC}$(Hz) for $\phi = 120°$ [c]	
	Normal	Cutoffs[d]	Normal	Cutoffs[d]
0	2.0	5.2	3.0	2.3
45	2.6		2.8	
90	3.5	5.7	2.3	2.0
135	3.1		2.2	
180	2.3	5.7	2.3	2.1
225	3.0		2.6	
270	3.5	5.7	2.5	2.1
315	2.6		2.4	

[a]Defined by the dihedral angle of C—C—C—OH, viewed clockwise from the terminally substituted carbon. That is, when $\phi = 0°$, the —OH group points toward the —CH₃ group.
[b]That is, coupling of the CO₂H carbon to the cis methyl.
[c]That is, coupling of the CO₂H carbon to the trans methyl.
[d]These calculations were performed with overlap integrals reduced to zero between the coupling atoms and all associated atoms, i.e., between CH₃— and —CO₂H.

from these compounds can be compared with those of compounds **6, 33, 8,** and **37.** It is regrettable that the extra bridge of compounds **61** and **64** distorts the overall molecule[3] but comparison of the data may be informative. First, compounds **62, 6,** and **33** are to be studied to compare the cis couplings in carboxyl systems. It is to be noted that the cis coupling in the lactone **62** (3.1 Hz) is larger than those in **6** and **33** (2.09 and 1.9 Hz, respectively). This observation is consistent with the predictions of Table 3-5, which suggest the coupling should be larger when the plane of the carboxyl group is perpendicular to the C—C—C—C framework. It is to be noted that this cis coupling in **62** is larger even when the twisting in this molecule should increase the dihedral angle of the cis C—C—C—C framework substantially (as much as $30°^3$), which should decrease the coupling owing to the Karplus-like relationship. Second, compounds **64**, **8**, and **37** are likewise studied to compare cis couplings in sp^3 systems. Here, the cis coupling is smaller in the distorted system **64** (2.8 Hz) than in the conformationally free systems **8** and **37** (4.07 and 4.57 Hz, respectively). Again, these results mirror the calculation of Table 3-4, which suggests the coupling should be larger when the hydroxyl group is free to swing away from the vicinal carbon. Third, in contrast to the opposite trends of cis carboxyl couplings and cis sp^3 systems noted immediately above, trans couplings follow consistent trends irrespective of the functionality. Thus, the couplings to the trans methyl in the distorted **61** (2.2 Hz) and **63** (2.6 Hz) are similar and different from the corresponding similar couplings in **6** (0.98 Hz) and **8** (0.61 Hz). Additionally, the couplings to the 7 position are always smaller in each distorted system than in the analogous couplings in the free systems: the couplings in **61** and **62** (4.9 and 3.7 Hz, respectively) are smaller than those in **6** and **33** (5.96 Hz and 5.2 Hz, respectively), and those in **63** and **64** (2.7 and 2.7 Hz, respectively) are smaller than those in **8** and **37** (3.64 Hz and 4.5 Hz, respectively). Therefore, the trans couplings appear more likely to reflect not differences in the orientation of the terminal substituent, but differences in the geometry of the C—C—C—C framework. An overall consistent picture thereby emerges where the Karplus-like relationship is paramount in determining $^3J_{CC}$, but where the orientation of the oxygen substituent can substantially affect cis couplings.

Another series of compounds that appears to reflect the effect of the orientation of the terminal substituent in cis couplings is **5–10**. In these compounds the trend is $^3J_{C-CO_2H} < {}^3J_{C-CHO} \approx {}^3J_{C-CH_2OH}$ (there is also a quite substantial trend where exo–exo couplings are larger than endo–endo couplings analogous to the same phenomenon in proton–proton couplings,[31] which obscures the effect of the terminal substituent: see Other Contributions to Carbon–Carbon Vicinal Couplings, this chapter). This trend is consistent with the calculations of Tables 3-3, 3-4, and 3-5. These tables predict that for the conformational free groups —CO_2H, —CH_2OH, and —CHO, where the functionality tends to rotate away from the vicinal carbon, the coupling should be greater for the —CH_2OH and —CHO groups. In the case of the —CO_2H group, of course, there are functional groups on both sides of the carbon atom, as discussed above. Accordingly, the vicinal cis coupling is smaller for the carboxylic acid, despite an additional electronegative terminal substituent, which ordinarily (all other things being equal) increases $^3J_{CC}$.

The role of the orientation of the hydroxyl group upon $^3J_{CC}$ is amply demonstrated in the series of cyclopropanes **50**, **52**, **53**, and **54**. As the

functionalized carbon becomes sterically encumbered by more methyl groups, the hydroxyl group is more likely to be oriented toward the vicinal cis methyl group and the $^3J_{CC}$ decreases through this series—from 3.4 to 1.04 Hz. The substituent effect of the methyl groups should be minor (for the effect of alkyl substituents upon $^3J_{CC}$, see the following section); in any case, the overriding nature of the orientation effect is clearly shown in the diastereomers **52** and **53** which show quite different J values (3.23 and 2.72 Hz, respectively) and yet whose only molecular difference is the relative orientation of groups in space. The trend in **50**, **52**, **53**, and **54** parallels the calculations of Table 3-3—$^3J_{CC}$ can vary over threefold for cis geometries, if a terminal hydroxyl group preferentially occupies different orientations.

This dependence of $^3J_{CC}$ upon the orientation of the terminal electronegative substituent has potentially serious consequences on the Karplus-like curves of Figures 3-1, 3-2, and 3-3. In these figures considerable spread exists along the ordinate axis at $\phi = 0°$ and at $\phi = 180°$. However, this spread is constant despite an expectation that it should be quite variable (i.e., small at $\phi = 0°$, moderate for —CO_2H at $\phi = 0°$, and large for —CH_2OH at $\phi = 0°$). Therefore, one can conclude that the complication of the terminal substituent conformation is not severe for the groups —CO_2H, —CH_2OH, and —CHO. Apparently these groups are in a relatively constant orientation to the coupling cis carbon, and substituent effects are overshadowed by other considerations (see next section).

The unimportance of the orientation of the terminal substituent on the data of Figures 3-1, 3-2, and 3-3 signifies that complications of this sort are not overriding considerations in conformationally free and unemcumbered systems. However, where the carbinol carbon is alkyl substituted (**50, 52, 53,** and **54**) and the hydroxyl group definitely changes its average orientation, the observed $^3J_{CC}$ values are quite different. Thus, the cis-$^3J_{CC}$ value of the tertiary alcohol **54** (1.04 Hz) would fall far below the range of the cis-$^3J_{CC}$ values of the primary alcohols of Figure 3-2 (ca. 3.4–5.8 Hz).

From all of the foregoing, one can conclude that insight into the orientation of a terminal hetero substituent can be obtained from values of $^3J_{CC}$ for cis couplings, and that this analysis is applicable to systems in which the hetero substituent is definitely restricted in its motion (e.g., in heterocyclic systems or in compounds with bulky groups). For conformationally free systems, as those represented by Figures 3-1, 3-2, and 3-3, the range of $^3J_{CC}$ at discrete values of ϕ must be due mainly to other factors. These additional considerations are discussed next.

Other Contributions to Carbon–Carbon Vicinal Couplings

In the classical view, two vicinal nuclei can couple by mechanistic contributions between orbitals about these nuclei and about intervening atoms in the formally bonded linkage.[32] Hence, one can consider coupling between the protons in H—C—C—H to be a summation of couplings via various pathways involving combinations of the hydrogen orbitals and the carbon orbitals (see Figure 3-5). This view appears to be overly simplistic, because it would lead one to conclude, for example, that the exo–exo and endo–endo H—H couplings in norbornane (**65**) are similar. However, the experimental $^3J_{HH}$ values are quite different—12.2 and 9.0 Hz, respectively.[31] It was shown by a theoretical study

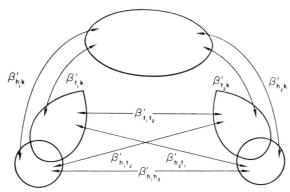

Figure 3-5. Individual mechanistic contributions to vicinal coupling. (Reproduced with permission.[32])

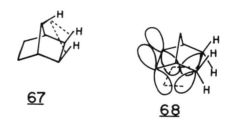

$$\underline{65} \qquad \underline{66}$$

12.2 (exo-exo)

9.0 (endo-endo)

9.30 (exo-exo)

9.02 (endo-endo)

that in fact other through-space interactions contributed quite significantly to the overall coupling, predominantly via the C-7 carbon and its hydrogens.[31] In this theoretical study, these interactions were artificially removed in the calculations to demonstrate a positive contribution to $^3J_{exo-exo}$ via the 7 bridge (see **67**). This method of removing interactions was by setting the appropriate overlap integrals to zero, and was the same method used as described in Tables 3-3, 3-4, and 3-5. The empirical results for norbornane (**65**) are to be compared with those for norbornene (**66**), whose exo–exo and endo–endo couplings are about the same—9.3 and 9.0 Hz, respectively.[31] In this study, it was shown that the near equivalence of these two $^3J_{HH}$ values for **66** was fortuitous. By removing the overlaps between the back lobes of the exo C—H bonds and the p orbitals of the olefins, it was seen that in **66** a negative contribution to $^3J_{exo-exo}$ is afforded by this "backdoor" route (see **68**). Thus, the

$$\underline{67} \qquad \underline{68}$$

observed exo–exo coupling in **66** is reduced by the mechanism shown in **68** to a value similar to the endo–endo coupling in **66**. It is disconcerting to realize that the near identical values of $^3J_{exo-exo}$ and $^3J_{endo-endo}$ in norbornene **66** do not necessarily reflect near identical geometries of the H—C—C—H coupling path. Instead, $^3J_{exo-exo} \approx {}^3J_{endo-endo}$ in **66** because of a particular combination of geometric considerations including not only the geometry of the formal bonding pathway H—C—C—H but also the positions of neighboring groups. Generalizations are difficult to make except to note that each system is unique and that vicinal proton–proton couplings indicate only approximately the involved dihedral angle. Vicinal proton–proton couplings can be powerful and convincing data in conformational analysis—for example, in distinguishing axial–axial and axial–equatorial relationships in cyclohexanes, which differ in dihedral angle by 120°. However, the limitations of this approach become obvious if one attempts to use $^3J_{HH}$ values to differentiate between similar H—C—C—H geometries, e.g., those with less than 50° difference.

The lesson afforded by **65** and **66** allows a possible explanation for the spread of carbon–carbon ^3J values near $\phi = 180°$ in Figures 3-1, 3-2, and 3-3. At first inspection it appears from these figures that near $\phi = 180°$, $^3J_{CC}$ actually decreases with an increase in the dihedral angle. However, it was noted that the compounds with $\phi \approx 180°$ are of special types, notably adamantanes, which are subject to particular through-space contributions.[12]

To illustrate dramatically the large number of through-space interactions even in such a simple molecule as butane, Table 3-6 was developed.[33] In this table theoretical calculations were performed on butane with various non-bonded interactions artificially removed, in the manner described above (except that the complete elements of the Fock matrices were set equal to zero instead of just the overlap integral; this new method is superior but does not change qualitative conclusions of the former method). In this table, butane is rotated through several selected values of the dihedral angle as the nonbonded interactions are removed. The first column of J values is normally calculated ones. Succeeding columns give calculated J values as specific interactions between various carbon and hydrogen orbitals are eliminated, and the listed values reflect the contribution provided by coupling along the nonbonded path indicated. For example, column A specifies the contribution between the orbitals of C-1 and C-3, and of C-2 and C-4; the contribution is negative and is significant even when $\phi = 180°$, signifying "rear lobe" interactions shown in **69**. Column B shows that when $\phi = 0°$, a through-space interaction between C-1

TABLE 3-6. CALCULATED INDO-FPT MO RESULTS FOR VICINAL ^{13}C—^{13}C COUPLINGS CONSTANTS IN BUTANE EXCLUDING SELECTED COUPLING PATHS AS A FUNCTION OF THE DIHEDRAL ANGLE $\phi^{a,b}$

Dihedral angle ϕ (degrees)	$^3J_{CC'}$	A $\Delta J_{CC'}{}^c$	B $\Delta J_{CC'}{}^d$	C $\Delta J_{CC'}{}^e$	D $\Delta J_{CC'}{}^f$	E $\Delta J_{CC'}{}^g$	F $\Delta J_{CC'}{}^h$	G $\Delta J_{CC'}{}^i$
0	5.79	−1.53	−1.05	6.43	−4.35	−0.35	5.43	0.40
30	3.96	−1.04	−0.89	4.40	−3.22	−0.32	4.10	0.38
60	1.97	0.05	−0.51	1.72	−0.79	−0.19	1.77	0.28
90	0.56	0.17	−0.26	0.03	0.22	−0.01	0.23	−0.01
120	1.45	−0.55	−0.16	−0.18	0.28	0.34	0.54	−0.26
150	3.34	−1.44	−0.17	−0.13	0.15	0.62	1.72	−0.36
180	4.27	−1.79	−0.19	−0.10	0.10	0.70	2.34	−0.36

^{a}All values are in hertz.
^{b}Reproduced with permission.[33]
^{c}All elements of the Fock matrices associated with the valence atomic orbitals connecting the C-1 and C-3 carbon atoms and connecting the C-2 and C-4 carbon atoms were set equal to zero in each self-consistent field (SCF) cycle ($*CH_3CH_2CH_2*CH_3$ and $*CH_3CH_2CH_2*CH_3$).
$^d*CH_3CH_2CH_2*CH_3$.
$^e*CH_3CH_2CH_2*CH_3$.
$^f*CH_3CH_2CH_2*CH_3$ and $*CH_3CH_2CH_2*CH_3$.
$^g*CH_3CH_2CH_2*CH_3$ and $*CH_3CH_2CH_2*CH_3$.
$^h*CH_3CH_2CH_2*CH_3$ and $*CH_3CH_2CH_2*CH_3$.
$^i*CH_3CH_2CH_2*CH_3$ and $*CH_3CH_2CH_2*CH_3$.

and C-4 is negative (see **70**). Through-space interactions can be very significant through hydrogens, also—column C shows that the contribution to J_{CC} is positive and large through the terminal hydrogens when $\phi = 0°$ (see **71**). The interactions in **71** are largely offset by terminal carbon–hydrogen interactions (column D; see **72**). Column E shows that the contribution between H-1 and H-3 (and H-2 and H-4) is small (see **73**). However, the contribution between H-1 and C-3 (and C-2 and H-4) is quite significant, particularly when $\phi = 0°$ (column F; see **74**). By contrast, interactions between C-1 and H-3 (and H-2 and C-4) are small (column G; see **75**).

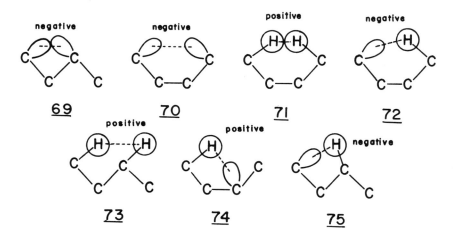

The complications embodied in Table 3-6 and structures **69–75** portend ominously of additional complexities as one moves to larger molecules. As hydrogens in **69–75** are substituted by alkyl groups, corresponding interactions are changed. In reality, only interactions **73** and **75** apply, because these hydrogens are internal. Substituting these hydrogens by alkyl substituents will alter the corresponding through-space contributions for a C—C—C—C framework of the same geometry. By contrast, interactions **71**, **72**, and **74** involve only terminal hydrogens. Substituting these terminal hydrogens by alkyl groups is not really possible, because an alkyl substituent on the terminal carbon will simply rotate away, exposing another terminal hydrogen to interact in the same manner. Therefore, for a set of calculations for alkyl-substituted butanes, one should be especially alert to C-2 (and C-3) substitutions for significant changes in $^3J_{CC}$. Such calculations are given in Table 3-7.[34] The entries in this table show the calculated carbon–carbon couplings as various methyl groups substitute the hydrogens. Inspection of these results makes it clear that modest, but significant, reductions in $^3J_{CC}$ occur when the C-2 (or C-3) hydrogen is methyl substituted. These reductions in $^3J_{CC}$ occur because the positive contribution in **73** (Table 3-6, column E) is being removed.

Also given in Table 3-7 are the calculated results for cyclohexanes and adamantanes, derivatives of which represent the "anomalously" low values near $\phi = 180°$ in Figures 3-1 and 3-2. Precisely as experimentally observed in these figures, the calculated $^3J_{CC}$ values in Table 3-7 are quite low for these compounds.

The effect of substituting the C-2 (C-3) hydrogen discussed in the previous two paragraphs has been termed the "γ-substituent effect".[11, 34] The dubbing of "γ" instead of "β" is because this effect is seen also in carbon–hydrogen couplings.[11, 34] Accordingly, not only is this phenomenon seen in 76 (J_{CC}) but

also in 77 (J_{CH}). Thus, just as $^3J_{CC}$ is small in 1-methyladamantane (21, Table 3-2), the geometrically equivalent C—H coupling (78) is also small[11] [compare 5.33 Hz for 78 with calculated value for $\phi = 180°$ of 8.81 Hz (Table 2-4) and with the experimental C—H coupling of 8.12 Hz for 79[35]].

The γ effect is no more dramatically seen than in a comparison of the $^3J_{CC}$ values near $\phi = 180°$ for 17, 19, 30, and 36 (taken from Table 3-2). The

hydrogen–hydrogen interaction responsible for the γ effect shown in 73 is apparently quite efficient in 30 and 36. In 30 and 36 the pertinent hydrogens (circled) are seen to locked rigidly in a conformation very close to one another. The result are "anomalously" large $^3J_{CC}$ values for 30 and 36. More direct experimental confirmation of the γ effect is afforded by compounds 6, 33, 12, 8, 37, and 14 (all taken from Table 3-2). In compounds 6, 33, 8, and 37 a γ hydrogen–hydrogen interaction is possible, geometrically the same as in 30 and 36. In compounds 12 and 14 the 7 hydrogen is substituted by a methyl group and no γ hydrogen–hydrogen interaction is possible. The result are substantially lower values of $^3J_{CC}$ in 12 and 14.

TABLE 3-7. CALCULATED INDO-FPT MO RESULTS FOR VICINAL COUPLING CONSTANTS $^3J_{CC'}(180°)$ IN BUTANE, METHYL-SUBSTITUTED DERIVATIVES, AND CYCLIC COMPOUNDS[a]

Molecule	$^3J_{CC}$(Hz)	Molecule	$^3J_{CC}$(Hz)
	4.27		2.76
	4.26		3.75
	4.11		4.30
	3.81		4.27
	3.44		4.47
	3.81		3.72
	3.44		3.32
	3.37		3.39

TABLE 3-7. (continued)

Molecule	$^3J_{CC}$(Hz)	Molecule	$^3J_{CC}$(Hz)
	3.40		
	3.06		3.32

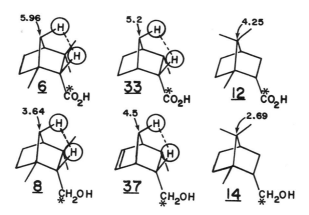

"Reproduced with permission.[34]

A different way to observe the same γ effect phenomenon is to compare $^3J_{CC}$ values in similar systems. Accordingly, one would expect the $^3J_{CC}$ values in the carboxylic acids **40–42** and in the alcohols **43–45** to be similar, because the same number and geometry of hydrogen–hydrogen interactions exist in each of these two classes of compounds. Indeed, comparison of the corresponding J values shows this is the case. For the compounds **40–42** the range of $^3J_{CC}$ values is 3.39–3.81 Hz, and for **43–45** the range is 3.09–3.51 Hz.[14] Thus, in each of the two types of compounds the range of $^3J_{CC}$ is less than 0.5 Hz. Inspection of Figures 3-1 and 3-2. puts these variations into proper perspective—they are quite small. These ranges of $^3J_{CC}$ are small, notwithstanding changes of hybridization at C-1 (the point of attachment of the —CO_2H and —CH_2OH groups). Therefore, the effect of the through-space interactions appears to be more important than molecular strain.

It is instructive to compare the range of $^3J_{CC}$ at $\phi = 180°$ observed above with $^3J_{HH}$ at $\phi = 180°$. Figures 3-1 and 3-2 show that with a trans geometry with $\phi = 180°$, the value of $^3J_{CC}$ may vary as much as 100%. For proton–proton couplings with a trans geometry with $\phi = 180°$, $^3J_{HH}$ may range 8–14 Hz,[36] i.e., as much as 75%. Hence, carbon–carbon couplings, with a greater number of through-space interactions have a greater potential range. Nevertheless, these through-space interactions are not so great as to prevent conformational analysis. One merely needs to remain vigilant to the various aspects of couplings and to be perhaps somewhat more careful than with proton–proton couplings. For one who is well versed in conformational analysis using proton–proton couplings, this habit of remaining cautious is already well ingrained and using carbon–carbon couplings for the same task should be no more worrisome.

Possible Complications Involving Internal Substituents

So far the topics of terminal substituents and of through-space interactions have been considered, but the effect of internal substituents upon $^3J_{CC}$ has not been discussed. In the fragment **80**, for example, the 2 substituent should bear an influence different from that of a 1 substituent (see $^3J_{CC}$ vs. Orientation of a Terminal Substituent, this chapter). Fortunately, this influence is not nearly as

$$\underset{\textbf{80}}{C-C-\overset{\overset{\displaystyle OH}{|}}{C}-C}$$

complicating a feature as it was for 1 substituents, because the 2 substituent is not free to rotate irrespective of the overall carbon framework. Thus, as the dihedral angle of C—C—C—C changes through the range 0° to 360°, the dihedral angle of C—C—C—O varies through the same range, either preceding

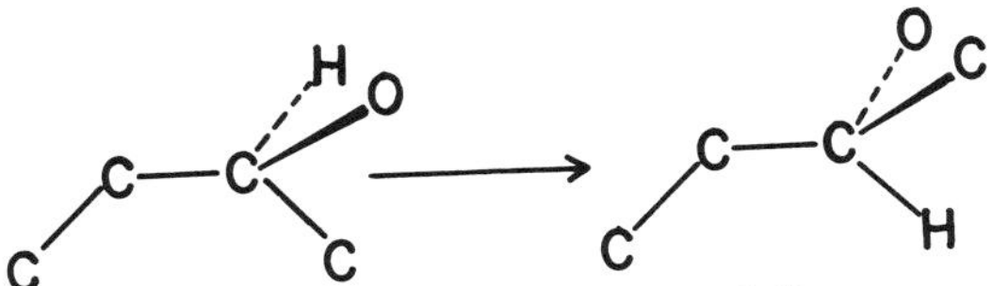

81   82    85

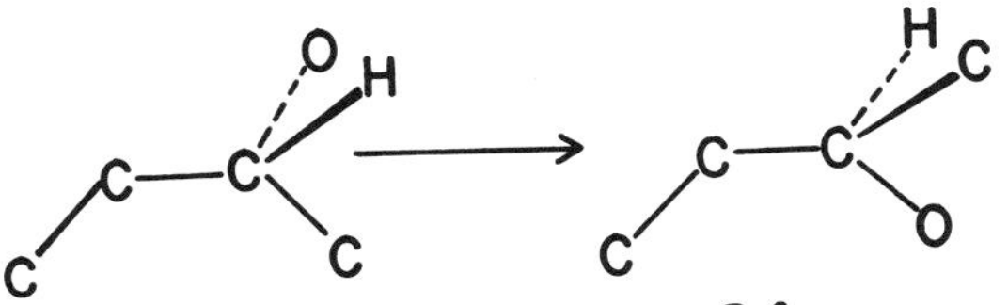

83   84 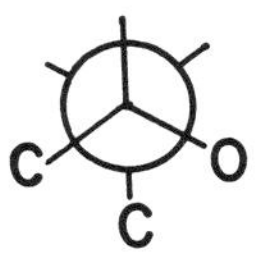   86

or lagging by 120° (depending upon the diastereomer; compare **81** and **82**, with **83** and **84**). The result is two possible dihedral angles for C—C—C—O for the same dihedral angle of C—C—C—C. For example, in both Newman projections **85** and **86** the C-1 and C-4 carbons are gauche but have different orientations with respect to the oxygen atom. To explore this potentially complicating feature, a study was conducted[15] wherein Table 3-8 was prepared, which includes calculated values of $^3J_{CC}$ of 2-butanol vs. the dihedral angle of

TABLE 3-8. CALCULATED VICINAL CARBON–CARBON COUPLING CONSTANTS OF BUTANE, 2-BUTANOL, AND BUTANOIC ACID VS. DIHEDRAL ANGLE OF THE C—C—C—C FRAMEWORK[a]

Dihedral angle, ϕ (degrees)	$^3J_{CC}$(Hz) of Compound		
	Butane	2-Butanol	Butanoic Acid
0	5.79	5.34	1.76[b]
30	3.96	3.79	2.27
60	1.94	1.97	1.24
90	0.56	0.50	0.71
120	1.45	1.10	2.14
150	3.34	2.77	4.65
180	4.27	3.82	5.87
210	3.34	3.31	4.65
240	1.45	1.57	2.14
270	0.56	0.51	0.71
300	1.94	1.56	1.24
330	3.96	3.40	2.27

[a]Reproduced with permission.[15,33]
[b]The —CO$_2$H group is coplanar with the C—C—C—C framework; see Table 3-5.

the framework. In this study, various ^{13}C-labeled compounds of the general structure **80** were prepared and the appropriate $^3J_{CC}$ values were measured. These $^3J_{CC}$ values were in substantial agreement with the calculated values of Table 3-8. Along with calculated $^3J_{CC}$ values for 2-butanol, Table 3-8 also lists corresponding $^3J_{CC}$ values for butane to allow ready analysis the effect of the 2 substituent. Also included are $^3J_{CC}$ values for butanoic acid to allow ready comparison of J values of a different type of substituent. Comparison of the $^3J_{CC}$ values for butane and 2-butanol makes it clear that for all dihedral angles of C—C—C—C, the effect of the 2-OH substituent is small and is generally attenuating. That an electronegative 2 substituent reduces $^3J_{CC}$ may merely reflect the general attenuating effect of such a substituent for carbon and proton couplings (see General Aspects of Aliphatic C—H Couplings, Chapter 2), or in addition may partly arise from the "γ" effects[34] by removing an interaction as depicted in **73**.

One can conclude that, in contrast to terminal substituents, internal substituents have few complicating features, and analysis of internally substituted butane moieties is relatively straightforward.

Conformational Analysis Using Vicinal Carbon–Carbon Couplings

To complete the tabulated information in Table 3-8, calculated results are presented in Table 3-9 for 1-butanol. The data of Table 3-9 complement those in Table 3-3, which gives $^3J_{CC}$ values only at $\phi = 0°$ and 120°. As already noted in Table 3-3, the $^3J_{CC}$ values of Table 3-9 do not depend much on the orientation of the —OH group except for dihedral angles $\phi \approx 0°$ of C—C—C—C.

The data of Tables 3-8 and 3-9 serve as the foundation for the major fundamental types of butane moities. Substitution of the OH group by other electronegative groups should have modest effect on the $^3J_{CC}$ values,[1,12] and accordingly the $^3J_{CC}$ values may be extrapolated to other functionalities (see Table 3-1). Armed with the principles discussed above regarding through-space interactions, one can proceed to actual cases. So far the number of applications is limited, but a few examples exist to illustrate the potential method.

Compound **87** appeared[9] as an isolated case that at the time only exemplified the problem of correlating C—C couplings with geometric structure. Now **87**

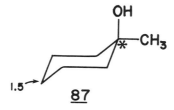

87

TABLE 3-9. INDO-FPT CALCULATED VICINAL
^{13}C—^{13}C COUPLING CONSTANTS FOR 1-BUTANOL FOR
VARIOUS VALUES OF ϕ AND ϕ' [a]

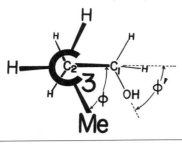

Dihedral angles		$^{3}J_{CC}$
ϕ	ϕ'	(Hz)
0	60	3.66
0	120	7.11
0	180	6.70
30	60	3.49
30	120	7.45
30	180	4.63
60	60	1.97
60	120	2.33
60	180	2.19
90	60	0.60
90	120	0.59
90	180	0.61
120	60	1.44
120	120	1.83
120	180	1.73
150	60	3.38
150	120	4.20
150	180	3.94
180	60	4.38
180	120	5.28
180	180	5.02

[a]Reproduced with permission from Wiley Heydon Ltd.[10]

can be analyzed as a 1-butanol moiety with $\phi = 60°$ and with no unusual interactions seen, for example, in the norbornane system. Table 3-9 suggests the indicated coupling should be ca. 2 Hz. Agreement of this value with the observed value of 1.5 Hz is quite reasonable.

An identical geometry of the C—C—C—C framework with $\phi = 60°$ exists for the adamantane compounds **91–95** (see Table 3-2). In these compounds[16] there exists again a terminal substituent which is OH, NH_2, $NH_3{}^+$, NHCOPh, and CO_2H. In each of these compounds two orientations of ϕ' (i.e., the dihedral angle of C—C—C—X) exist. Table 3-9 suggests that nevertheless the two $^3J_{CC}$ values for each compound should be similar. The experimental values (Table 3-2) bear this out for compounds **91–95** ($^3J_{CC}$ generally falls in the range 1.3–1.7 Hz). Apparent exceptions are the small $^3J_{CC}$ values for C-7 of **94** and **95**, but here through-space contributions are possible through the larger substituent. In fact, these considerations suggest that the original assignments[16] for C-5 and C-7 in **94** and **95** should be reversed. These reversals are reflected in Table 3-2.

In compounds **101–103**[17] the six-membered ring is conformationally free and C—C couplings may indicate the conformation of this ring. The small couplings of C-6 to C-2 are not surprising in any conformation, because the dihedral angle ϕ is rigidly locked and is approximately 90°. The coupling of C-6 to C-3 is small, also, in a C—C—C—C linkage that is conformationally mobile. If the conformation of the six-membered ring were a true chair, then $\phi \approx 45°$ and Figures 3-2 and 3-3 would suggest much higher $^3J_{CC}$ values than those observed. The conclusion is that the six-membered ring is actually flattened[17] and is more like the conformation depicted in **88**. The small $^3J_{CC}$ values also would be in accord with a boat conformation for the six-membered ring ($\phi \approx 115°$), but it is highly improbable that this is the exclusive conformation for the series **101–103**.

Compound **55** (from Table 3-2) is an example of a system whose $^3J_{CC}$ values help considerably in elucidating the conformation of a cyclic system. The four

couplings to the vicinal carbons are as shown[8] (assignments were made with the help of lanthanide-shift and selective proton irradiation studies). The first item of note is the small coupling to both methyl groups (C-7 and C-8). If the C-6 carbon were eclipsed, or were antiperiplanar, to either of these carbons, the couplings should be much larger (from Table 3-8). The conclusion is that the labeled carbon is roughly staggered between the two vicinal carbons. Also, the values of $^3J_{CC}$ to C-3 and C-4 are too small for a dihedral angle near 180°. Study of a model strongly suggests the conformation **89** complies with these restrictions. In the conformation **89** the cyclopentane molecule is in an envelope conformation with the tab holding the OH group to be folded such that the methyl group is pseudoequatorial. (To see these features clearly it is suggested that the reader construct a model.) If the molecule is puckered too greatly, the dihedral angles involving C-3 and C-4 approach 180° and additionally C-6 bisects the (C-8)–(C-2)–(C-7) angle. If, however, the ring is not so strongly puckered, the dihedral angles involving C-3 and C-4 recede from 180° and the cis dihedral angle involving C-6 and C-7 approaches 0°. Indeed, this cis $^3J_{CC}$ ($J_{6,7}$) is greater than the trans $^3J_{CC}$ value ($J_{6,8}$), suggesting this geometry. The data are therefore strongly suggestive of a rather specific conformation for **55** in which the C-6 methyl is rotated downward. This overall conformation (**89**) appears to be reasonable; the C-8 and C-9 methyl groups twist away from one another as partial staggering occurs between C-1 and C-2 and their substituents.

Compound **57** is another cyclopentane whose C—C couplings aid in conformational analysis (taken from Table 3-2). In this molecule the $^3J_{CC}$ value

involving the C-8 methyl group is large (i.e., large for a —*CO_2H group; see Table 3-8 and Figure 3-1) and indicates a dihedral angle $\phi \approx 0°$. The $^3J_{CC}$ values to C-4 and C-3 are smaller and suggest dihedral angles far from 180° (it is to be remembered that for carboxylic acids $^3J_{CC}$ for $\phi = 0°$ is smaller than $^3J_{CC}$ for $\phi = 180°$, in contrast to couplings for alkanes, such as **55**, where the reverse is the case). It is again suggested that the reader use models to see these geometrical considerations clearly. The conclusion to be reached from these geometrical restrictions is that **57** may be slightly puckered with the C-7 methyl slightly tending toward a pseudoequatorial orientation (see **90**).

In compound **60** the methyl group is locked into the equatorial orientation. The vicinal C—C coupling between the methyl group and the C-3 carbon is 3.2 Hz,[15] somewhat smaller than the calculated value of 3.82 Hz from the model compound 2-butanol ($\phi = 180°$, Table 3-8), but reasonable in view of the attenuating effects in cyclohexanes (see Other Contributions to Carbon–Carbon Vicinal Couplings, this chapter). It would be interesting to observe the behavior of a mobile cyclohexane whose conformations could be "frozen out," and indeed a low-temperature study of cis-1,4-dimethylcyclohexane (**180**) has been conducted.[27] The two conformers **180a** and **180b** (with the labeled methyl group equatorial and axial, respectively) were observed independently in the NMR spectrum at 180°K. As anticipated from the dihedral angle relationship, the $^3J_{CC}$ value was larger in the equatorial isomer **180a** (compare 3.42 vs. ≤ 1 Hz). Contrastingly, the analogous $^3J_{CC}$ value in trans-1,4-dimethylcyclohexane (**181**) is 4.52 Hz, substantially larger than the 3.42 Hz value. It is difficult to explain this discrepency in terms of an "orientation"[27] effect; instead, a flattening of the cyclohexane ring[27] is more plausible. This latter interpretation has significant implications on basic working assumptions that the main role of cyclohexane substituents is to partition conformations without significantly affecting the classical symmetrical chair geometry.

The few examples given here attest to the capability of C—C couplings to aid in conformational analysis in a unique manner. The conformational analysis of cyclopentanes has lagged behind that of cyclohexanes principally because of the difficulties encountered in the study of a conformationally mobile five-membered ring. We have seen that C—C couplings appear to be a powerful complement to proton NMR. Now that the method of accurately measuring C—C couplings is available and has been developed for conformational analysis, additional systems may be investigated in an effort to obtain a greater understanding of their molecular geometries.[37, 38]

References

1. Marshall, J. L.; Miiller, D. E.; Conn, S. A.; Seiwell, R.; Ihrig, A. M. *Acc. Chem. Res.*, **7**, 333 (1974).
2. Sternhell, S. *Quart. Rev.*, **23**, 236 (1969).
3. Muntz, R. L.; Pirkle, H.; Paul, I. C. *J. Chem. Soc. Perkin Trans. 2*, 483 (1972); Altona, C.; Sundaralingam, M. *Acta. Cryst. A*, **25**, 5141 (1969); Altona, C.; Sundaralingam, M. *Acta. Cryst.*

B, **28**, 1806 (1972); Moriarty, R. M.; Gopal, H.; Flippen, J. L.; Karle, J. *Tetrahedron Lett.*, 351 (1972); Moews, P. C.; Knox, J. R.; Vaughan, W. R. *Tetrahedron Lett.*, 359 (1977); Brueckner, D. A.; Hamor, T. A.; Robertson, J. M.; Sim, G. A. *J. Chem. Soc.*, 799 (1962); Ferguson, G.; Fritchie, C. J.; Robertson, J. M.; Sim, G. A. *J. Chem. Soc.*, 1976 (1961); Fratini, A. V.; Britts, K.; Karle, I. L. *J. Phys. Chem.*, **71**, 2482 (1967); Cser, F.; Sasvari, K. *J. Chem. Soc. Perkin Trans. 2* 478 (1976); Chapuis, G.; Zalkin, A.; Templeton, D. H. *Acta. Cryst.* *B*, **29**, 2642 (1973); Chapuis, G.; Zalkin, A.; Templeton, D. H. *Acta. Cryst.* *B*, **33**, 560 (1977); Palmer, K. J.; Wong, R. Y.; Lundin, R. E.; Khalifa, S.; Casida, J. E. *J. Am. Chem. Soc.*, **97**, 408 (1975); Chapuis, G.; Zalkin, A.; Templeton, D. H. *Acta. Cryst.* *B*, **33**, 1290 (1977); Moutin, M.; Rassat, A.; Bordeaux, D.; Lajerowicz-Bonnetau, J. *J. Mol. Struct.*, **31**, 275 (1976); Flippen, J. L. *Acta. Cryst.* *B*, **28**, 2046 (1972); Ekejiuba, I. O. C.; Hallam, H. E. *J. Mol. Struct.*, **6**, 341 (1970); Altona, C.; Buys, H. R.; Havinga, F. *Rec. Trav. Chim.*, **85**, 973 (1966); Eliel, E. L.; Allinger, N. L.; Angyal, S. J.; Morrison, G. A. "Conformational Analysis"; Wiley-Interscience: New York, 1965, p. 441; Armitage, B. J.; Kenner, G. W.; Robinson, M. J. T. *Tetrahedron*, **20**, 747 (1964); MacDonald, A. C.; Trotter, J. *Acta Cryst.*, **19**, 456 (1965); Wing, R. M.; Tustin, G. C.; Okamura, W. H. *J. Am. Chem. Soc.*, **92**, 1935 (1970); Wilcox, C. F. Jr. *J. Am. Chem. Soc.*, **82**, 414 (1960); Altona, C.; Sundaralingam, M. *J. Am. Chem. Soc.*, **92**, 1995 (1970).

4. Marshall, J. L.; Faehl, L. G.; Ihrig, A. M.; Barfield, M. *J. Am. Chem. Soc.*, **98**, 3406 (1976).

5. Wasylishen, R.; Schaefer, T. *Can. J. Chem.*, **51**, 961 (1973).

6. Marshall, J. L.; Faehl, L. G.; McDaniel, C. R. Jr.; Ledford, N. D. *J. Am. Chem. Soc.*, **99**, 321 (1977).

7. Garbisch, E. W. Jr. *J. Am. Chem. Soc.*, **86**, 5561 (1964).

8. Barfield, M.; Marshall, J. L. Unpublished results.

9. Marshall, J. L.; Miiller, D. E. *J. Am. Chem. Soc.*, **95**, 8305 (1973).

10. Marshall, J. L.; Conn, S. A.; Barfield, M. *Org. Magn. Reson.*, **9**, 404, (1977).

11. Barfield, M.; Marshall, J. L.; Canada, E. D. Jr.; Willcott, M. R. III. *J. Am. Chem. Soc.*, **100**, 7075 (1978).

12. Barfield, M.; Conn, S. A.; Marshall, J. L.; Miiller, D. E. *J. Am. Chem. Soc.*, **98**, 6253 (1976).

13. Marshall, J. L.; Canada, E. D. Jr. *J. Org. Chem.*, **45**, 3123 (1980).

14. Barfield, M.; Brown, S. E.; Canada, E. D. Jr.; Ledford, N. D.; Marshall, J. L.; Walter, S. R.; Yakali, E. *J. Am. Chem. Soc.*, **102**, 3355 (1980).

15. Doddrell, D.; Burfitt, I.; Grutzner, J. B.; Barfield, M. *J. Am. Chem. Soc.*, **96**, 1241 (1974).

16. Berger, S.; Zeller, K.-P. *Chem. Commun.*, 649 (1976).

17. Berger, S. *J. Org. Chem.*, **43**, 209 (1978).

18. Chaloner, P. A. *J. Chem. Soc. Perkin Trans. 2*, 1028 (1980).

19. Berger, S. *Org. Magn. Reson.*, **14**, 65 (1980).

20. Stöcker, M.; Klessinger, M. *Org. Magn. Reson.*, **12**, 107 (1979).

21. Jokisaari, *Org. Magn. Reson.*, **11**, 157 (1978).

22. Weigert, F. J.; Roberts, J. D. *J. Am. Chem. Soc.*, **94**, 6021 (1972).

23. Dreeskamp, H.; Hildenbrand, K.; Pfisterer, G. *Mol. Phys.*, **17**, 429 (1969).

24. Ziessow, D. *J. Chem. Phys.*, **55**, 984 (1971).

25. Bax, A.; Freeman, R.; Kempsell, S. P. *J. Magn. Reson.*, **41**, 349 (1980).

26. Bax, A.; Freeman, R.; Kempsell, S. P. *J. Am. Chem. Soc.*, **102**, 4849 (1980).

27. Booth, H.; Everett, J. R. *Can. J. Chem.*, **58**, 2709 (1980).

28. Della, E. W.; Pigou, P. E. *J. Am. Chem. Soc.*, **104**, 862 (1982).

29. Marshall, J. L.; Faehl, L. G.; Kattner, R. *Org. Magn. Reson.*, **12**, 163 (1979).

30. Barfield, M.; Dean, A. M.; Fallick, C. J.; Spear, R. J.; Sternhell, S.; Westerman, P. W. *J. Am. Chem. Soc.*, **97**, 1482 (1975).

31. Marshall, J. L.; Walter, S. R.; Barfield, M.; Marchand, A. P.; Marchand, N. W.; Segre, A. L. *Tetrahedron*, **32**, 537 (1976).

32. Murrell, J. N. In "Progress in Nuclear Magnetic Resonance Spectroscopy," Emsley, J. W.; Feeney, J.; Sutcliffe, L. H. Eds.; Pergamon Press: New York, 1970; Vol. 6, p. 34.

33. Barfield, M. *J. Am. Chem. Soc.*, **102**, 1 (1980).

34. Barfield, M.; Marshall, J. L.; Canada, E. D. Jr. *J. Am. Chem. Soc.*, **102**, 7 (1980).

35. Chertkov, V. A.; Sergeyev, N. M. *J. Am. Chem. Soc.*, **99**, 6750 (1977).

36. Bovey, F. A. "Nuclear Magnetic Resonance Spectroscopy"; Academic Press: New York, 1969, pp. 132–140.

37. For example, a possible extension of the use of carbon–carbon couplings in conformational analysis and chemical shift assignments may be in the field of organometallic compounds: preliminary results indicate stereochemistry can be determined in metal carbonyl compounds using $^2J_{CC}$ values. Aime, S.; Osella, D. *Chem. Commun.*, 300 (1981).

38. A recent example utilizing the double-quantum coherence technique (see Chapter 1) has involved the study of unlabeled l-octanol. All vicinal carbon–carbon couplings were revealed and the gauche/trans ratios about the carbon–carbon bonds were given. This study was conducted over a broad temperature range and allowed the calculation of the gauche–trans enthalpy differences for the bonds. Phillipi, M. A.; Wiersema, R. J.; Brainard, J. R.; London, R. E. *J. Am. Chem. Soc.*, in press.

4

CARBON–CARBON COUPLINGS IN π SYSTEMS

Carbon–Carbon Couplings: General

Carbon–carbon couplings in π systems can be separated into two groups, each with different coupling behavior: (1) systems wherein at least one of the coupled carbons is not in an olefin or aromatic group, such as **1**, **4**, and **19**, and

(2) systems wherein both coupled carbons and the entire coupling route are inherent in a conjugated system, such as **6**, **66**, and **72**.

In general, the former type of coupling follows a pattern quite similar to that in aliphatic couplings, i.e., geminal couplings (^2J) can be small (0–3 Hz), vicinal couplings (^3J) are about the same as for flat ($\phi = 0°$, 180°) aliphatic systems (ca. 3–5 Hz), and longer range couplings are small (<1 Hz). Examples below illustrate this behavior. The geminal couplings are 0.8–2.0 Hz in **2**, 3.10 Hz in **19**, and 1.65–3.74 Hz in **55**. Vicinal couplings, like aliphatic couplings, are generally larger: 5.8 Hz in **2**, 3.84 Hz in **19**, and 0.81–4.94 Hz in **55**. Lastly, longer range

couplings are quite small: 0–0.86 Hz in **19** and **55**. Our understanding of aliphatic couplings (Chapter 3) immediately permits an explanation for the small ^3J value of 0.81 Hz to the 8 position in **55**—through-space contributions via the carbonyl—and further strengthens the analogy between aliphatic couplings and this first group of π systems. It is to be further noted that the carbonyl group, although conjugated fully within the coupling route as in **55**, still behaves more like a substituent, as in aliphatic systems. In summary, there appear to be few surprises in moving from aliphatic systems to this first group of π systems beyond a few refinements: principally that couplings in the π system tend to be a little larger, a fact that is not surprising owing to the greater s character and shorter bonds associated with the coupling path. (A notably unique case is acetylenes, whose geminal couplings are quite large and whose vicinal couplings are small.) Couplings belonging to this first category have been studied in a variety of systems, including olefins, acetylenes, benzenes, naphthalenes, anthracenes, biphenyls, and [2.2]paracyclophanes. Examples from the literature of this type of coupling are found in Tables 4-1, 4-2, 4-3, 4-7, 4-8, 4-10, 4-11, 4-12, and 4-13 and, as noted before, in Table 3-2 (compounds **15** and **16**).

In contrast to the types of systems briefly discussed above, couplings where the entire coupling path is olefinic or aromatic depart more noticeably from aliphatic behavior. For example, in butadiene (**6**) the vicinal coupling is much

larger (9.05 Hz),[4] attesting to the significant effect of a conjugated system. Such a large ^3J value in **6** suggests possible π contributions to the transmission of the spin information from one nucleus to another. In other examples, toluene (**65**)[18] and anthracene (**71**),[17] the vicinal couplings are likewise large (3.13–9.45 Hz), with a variability that suggests the importance of cis vs. trans coupling paths and π-bond order. Another notable feature of π systems in this second group is the longer range couplings, such as the ^4J value of 1.62 Hz in anthracene (**71**). Pyrene (**84**)[20] abounds with examples of longer range couplings—there exists a five-bonded coupling of 2.27 Hz and even a six-bonded coupling is substantial (0.76 Hz)! Such long-range coupling naturally suggests efficient transmission via π bonds. Comparison of couplings in **84** with those in **87**[23] nicely illustrates the difference in the same system between the two broad categories of π systems. In **84** (second category), the longer range couplings are all significant and can be quite large; in **87**, wherein the coupling includes a substituent (first category), the couplings are well attenuated.

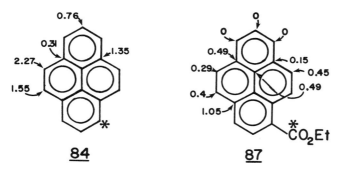

84   87

Finally, in this second category of π systems, geminal couplings remain consistent to the end—they are small and fall in a range similar to those in the first category. For example, in compounds **6, 65,** and **71** the values of ^2J are <1–1.80 Hz. However, in contrast to the trends of vicinal and longer range couplings in the second category of π systems (which tend to be larger than those in the first category), the geminal couplings in the second category are generally attenuated somewhat. Tables 4-4, 4-5, 4-6, 4-7, 4-9, and 4-12 include couplings in this second category of π couplings in naphthalenes, anthracenes, phenanthrenes, pyrenes, and methylene benzocycloalkenes, with a sole example of an olefin (**6**) in Table 4-1. In some instances signs of J values are listed (Tables 4-4 through 4-9), which were determined by means of special irradiation studies (compounds **16** and **17** in Table 4-1 also include signs). The significance of these signs is discussed in this chapter, but the description of the technique of sign determination is deferred to Chapter 5, Special Topics.

TABLE 4-1. CARBON-CARBON COUPLING CONSTANTS (AND CARBON-13 CHEMICAL SHIFTS) OF OLEFINS AND ACETYLENES

J_{CC}(Hz) to labeled carbon (δ_{C-13}, ppm) of carbon no.:

Compound	1	2	3	4	5	6	7	8	9	10	Other	References
	4.0 (14.6)	0.0 (21.3)	73.1 (127.2)	(138.5)	41.6 (32.5)	2.0 (21.6)	3.6 (13.7)	43.0 (39.7)	2.3 (22.0)	3.8 (13.9)		1
	(133.9)	73.6 (121.7)	0.8 (25.8)	5.8 (23.0)	2.0 (23.6)	40.1 (30.5)	44.8 (24.0)					1
	(139.6)	72.0 (123.9)	4.88 (32.6)	1.71 (23.9)	38.1 (36.8)	44.7 (16.4)						2
		(121.8)	0.0 (157.2)	2.38 (15.5)								3

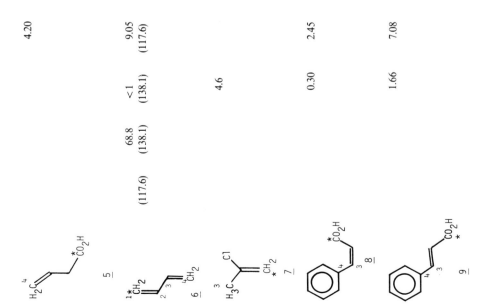

4.20				3
9.05 (117.6)	<1 (138.1)	68.8 (138.1)	(117.6)	4 (C-2, C-3) 53.70
	4.6			5
2.45	0.30			6
7.08	1.66			6

127

TABLE 4-1. (continued)

J_{CC}(Hz) to labeled carbon (δ_{C-13}, ppm) of carbon no.:

Compound	1	2	3	4	5	6	7	8	9	10	Other	References
10		(119.8)	1.30 (147.0)	7.30 (17.8)								3, 6
11			2.17	1.54	7.50							6
12			9.07									6
13	55.5 (42.2)	2.8 (122.2)	3.2 (126.2)	0.0 (26.4)								7

128

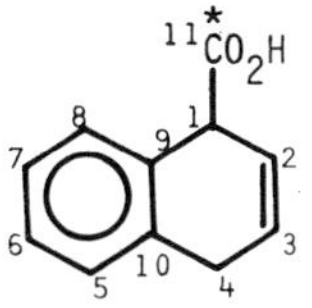 14	54.7 (47.2)	3.8 (124.2)	3.3 (128.0)	0.8 (30.3)	0.0 (128.9)	0.5 (127.6)	0.0 (126.7)	1.2 (129.7)	2.5 (123.9)	1.9 (134.9)		7
$\underset{3}{CH_3}\underset{2}{C} \equiv \overset{*}{\underset{1}{C}}H$ 15			11.8									8
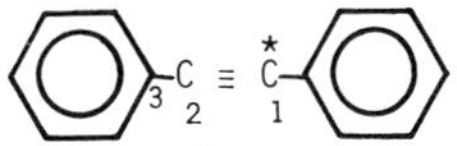 16			+13.1									9
$\underset{4}{CH_3}\underset{3}{C} \equiv \overset{*}{\underset{2}{C}}\underset{1}{C}O_2\underset{5}{C}H_3$ 17		+127.5	+20.3	+1.84	−2.28						(C-3, C-4) +65.15	10, 11
$CH_3O_2\underset{3}{C}\underset{2}{C} \equiv \overset{*}{\underset{1}{C}}CO_2CH_3$ 18			18.42									6

TABLE 4-2. CARBON-CARBON COUPLING CONSTANTS (AND CARBON-13 CHEMICAL SHIFTS) OF α-LABELED BENZENE DERIVATIVES

J_{CC}(Hz) to labeled carbon (δ_{C-13}, ppm) of carbon no.:

Compound	1	2	3	4	5	6	7	8	9	10	Other	References
*CH₃—phenyl 19	44.19 (137.8)	3.10 (129.1)	3.84 (128.5)	0.86 (125.4)								12
*CH₂OH—phenyl 20	47.72 (141.5)	3.45 (127.1)	3.95 (128.5)	0.73 (127.3)								12
*CH₂Cl—phenyl 21	47.78 (137.8)	3.69 (128.7)	4.23 (128.8)	0.69 (128.5)								12
*CO₂H—phenyl 22	71.87 (130.9)	2.54 (130.1)	4.53 (128.4)	0.90 (133.3)								12

130

Compound	1	2	3	4	Other	Ref
$CO_2^-Na^+$ **23**	65.90 (136.1)	2.23 (129.3)	4.11 (128.3)	0.8 (131.3)		12
CO_2CH_3 **24**	74.79 (130.6)	+2.38 (129.7)	+4.56 (128.5)	0.90 (132.9)	(CH_3) 2.63	10, 12
CHO **25**	53.2	3.98	4.61	1.14		13
$O=C-CH_3$ **26**	52.5	2.87	3.99	1.06	42.8	13
COCl **27**	74.35 (133.1)	3.53 (131.4)	5.46 (129.1)	1.18 (135.5)		12

TABLE 4-2. (continued)

J_{CC}(Hz) to be labeled carbon (δ_{C-13}, ppm) of carbon no.:

Compound	1	2	3	4	5	6	7	8	9	10	Other	References
28 O=C-Ph	54.8 (137.8)	2.71 (130.0)	4.03 (128.4)	2.03 (132.2)			(194.8)					13
29 CN	80.40 (112.2)	2.61 (132.1)	5.75 (129.2)	1.59 (132.9)								12
30 CO₂H / CH₃	72.4 (128.1)	2.62 (129.4)	4.48 (129.1)	0.90 (143.0)			(167.4)					13
31 CO₂H / Me	72.0 (129.4)	2.81 (130.8)	4.55 (138.3)	0.86 (134.6)	4.29 (128.4)	2.50 (127.4)	(172.8)					13

										Ref.
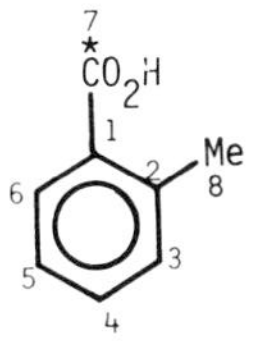 **32**	72.0 (129.2)	3.20 (140.8)	4.75 (131.9)	0.97 (132.5)	4.38 (125.9)	2.38 (131.4)	(170.3)	0.71 (21.9)		3, 13
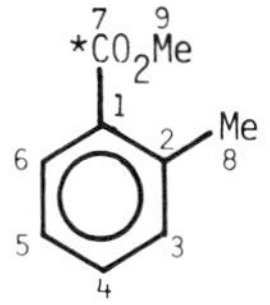 **33**	74.6 (129.2)	3.15 (140.2)	4.88 (131.8)	1.02 (131.9)	4.41 (125.8)	2.06 (130.7)	(166.9)	0.75 (21.69)	(51.3)	13
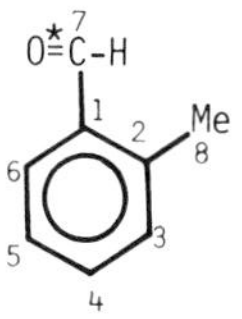 **34**	(134.8)	3.57 (140.6)	4.35 (132.2)	1.03 (133.8)	4.68 (126.7)	3.14 (132.2)	(192.6)	1.88 (19.4)		3
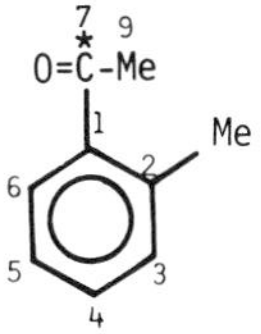 **35**	52.7 (137.8)	1.86 (138.3)	3.85 (129.5)	0.99 (131.4)	4.30 (125.8)	3.57 (132.0)	(199.2)	0.91 (21.5)	42.0 (28.9)	3, 13

Structures:

32: benzene ring with CO₂H group (positions labeled 7, ★, CO₂H, 1) and Me (8) at position 2; ring positions 1–6.

33: benzene ring with *CO₂Me (7, 9) and Me (8) at position 2.

34: benzene ring with O=C–H (7) and Me (8) at position 2.

35: benzene ring with O=C–Me (7, 9) and Me at position 2.

133

TABLE 4-2. (continued)

J_{CC}(Hz) to be labeled carbon (δ_{C-13}, ppm) of carbon no.:

Compound	1	2	3	4	5	6	7	8	9	10	Other	References
7 9 10 O=C–CMe$_3$ (ring with positions 1,2,3,4; Me$_8$ at 4) **36**	51.42 (135.5)	2.43 (128.3)	3.76 (128.7)	0.98 (141.4)			(208.2)	0.35 (21.4)	40.74 (44.1)	0.37 (28.2)		13
7 9 10 O=C–CMe$_3$ (ring with positions 1,2,3,4,5,6; Me$_8$ at 3) **37**	50.7 (138.9)	2.11 (128.4)	3.62 (137.8)	0.95 (131.5)	3.87 (127.8)	2.36 (124.8)	(172.6)	(28.05)		(29.7)		13
7 9 10 O=C–CMe$_3$ (ring with positions 1,2,3,4,5,6; Me$_8$ at 2) **38**	50.6 (137.6)	1.47 (134.0)	3.31 (130.7)	0.85 (128.5)	3.73 (124.9)	2.30 (124.6)	(211.7)	1.33 (19.7)	39.6 (44.5)	0.0 (28.0)		13

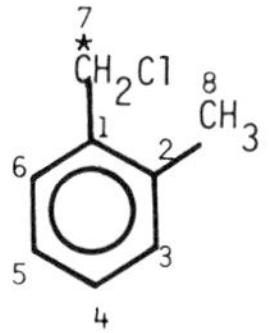 48.09 (134.9)	2.81 (136.4)	4.39 (130.0)	0.61 (128.2)	4.39 (125.7)	2.15 (129.1)	(44.3)	3.05 (18.5)	
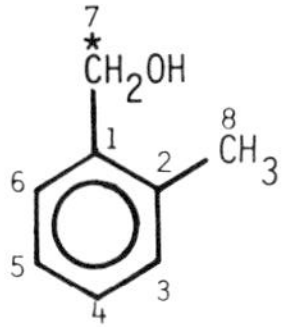 47.50 (139.7)	3.22 (135.9)	3.08 (130.2)	0.73 (127.4)	3.66 (126.1)	2.54 (127.6)	(62.5)	3.15 (18.5)	
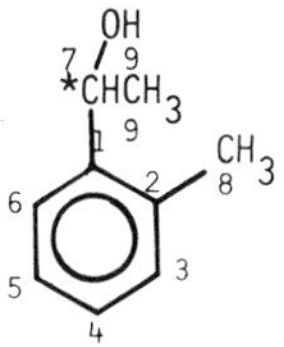 47.31 (145.0)	3.66 (133.7)	2.92 (129.9)	<0.5 (126.4)	3.05 (125.9)	1.34 (124.8)	(66.1)	2.82 (18.8)	38.15 (24.6)
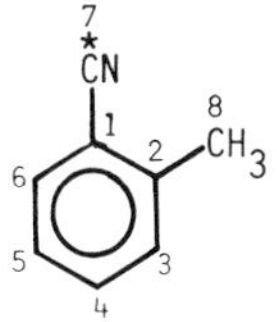 (117.4)	2.34 (141.0)	4.88 (129.7)	1.22 (132.1)	5.34 (125.8)	2.59 (131.8)	(112.3)	1.95 (20.2)	

39

40

41

42

3

3

3

3

135

TABLE 4-2. (continued)

J_{CC}(Hz) to be labeled carbon (δ_{C-13}, ppm) of carbon no.:

Compound	1	2	3	4	5	6	7	8	9	10	Other	References
Ph–CCl₂ (43)	51.6 (144.1)	3.21 (127.5)	4.52 (128.2)	1.02 (129.1)			(92.1)					14
CH₃, NO₂ (44)	43.55	~1.0	1.53	0.50	3.60	2.20						15
CH₃, NH₂ (45)	45.00	~1.0	1.64	0.62	3.87	2.47						15
CH₃, I (46)	46.85	<0.5	2.42	0.66	3.55	3.48						15

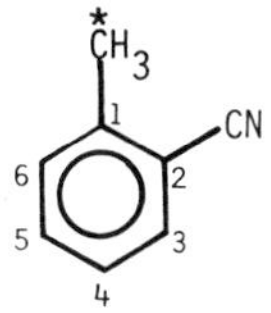

43.82 ~0.8 2.40 0.73 3.87 2.83 2.0 15

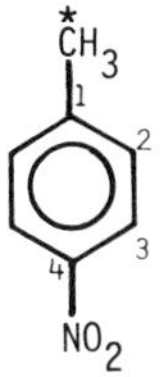

*CH_3 — CN — **47**

43.45 3.46 3.87 15

*CH_3 — NO_2 — **48**

45.91 3.17 4.19 <0.6 15

*CH_3 — NH_2 — **49**

+72.6 2.0 +4.1 −1.0 +4.5 +1.6 1.4 16

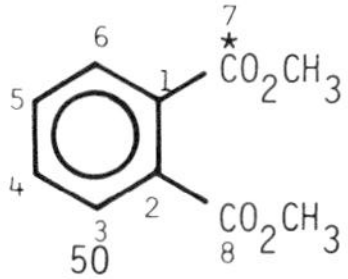

*CO_2CH_3 CO_2CH_3 — **50**

137

TABLE 4-2. (continued)

J_{CC}(Hz) to be labeled carbon ($\delta_{C\text{-}13}$, ppm) of carbon no.:

Compound	1	2	3	4	5	6	7	8	9	10	Other	References
51	+69.9	+2.6	+4.6	−1.0	+5.0	+2.8		5.6				16, 17
52				−1.0	+4.2			6.2				17
53		+3.49	+4.1	−0.9	+4.5	+2.5		8.8				17

Couplings in Nonconjugated Systems

Olefins and Acetylenes

Coupling patterns for olefins and acetylenes follow regular patterns. First, couplings in hydrocarbon olefins are quite similar to couplings in aliphatics. In **1**[1] the vicinal couplings appear "normal" in the range of 3.6–4.0 Hz, whether the coupling proceeds through the olefin or not. In **2** the vicinal coupling is larger (5.8 Hz) but the existence of a dual coupling path may account for this larger value.[2] Geminal couplings are near zero if the couplings are transmitted through the olefin (^2J is 0 Hz to C-2 in **1**, and 0.8 Hz to C-3 in **2**) and are somewhat larger

if not transmitted throughout the olefin (^2J is 2.0 Hz to C-5 in **2**[1]). Departures from these trends in ^2J may occur if the olefin is conjugated with an aromatic ring (see Table 4-12).

TABLE 4-3. CARBON-CARBON COUPLING CONSTANTS (AND CARBON-13 CHEMICAL SHIFTS) OF EXTERNALLY LABELED NAPHTHALENES

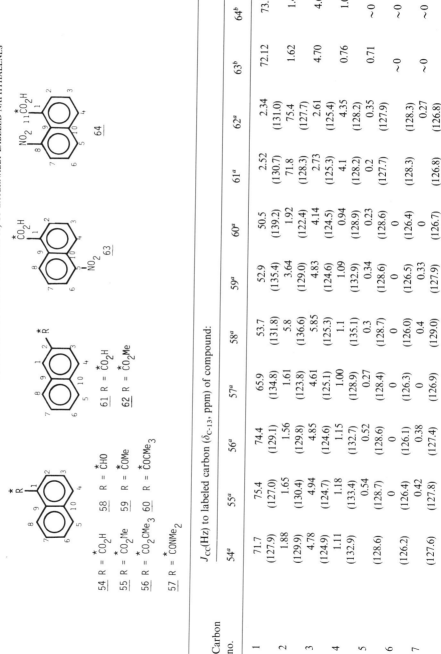

54   R = $^*CO_2H$
55   R = $^*CO_2Me$
56   R = $^*CO_2CMe_3$
57   R = $^*CONMe_2$

58   R = $^*CHO$
59   R = $^*COMe$
60   R = $^*COCMe_3$

61   R = $^*CO_2H$
62   R = $^*CO_2Me$

J_{CC}(Hz) to labeled carbon (δ_{C-13}, ppm) of compound:

Carbon no.	54^a	55^a	56^a	57^a	58^a	59^a	60^a	61^a	62^a	63^b	64^b
1	71.7 (127.9)	75.4 (127.0)	74.4 (129.1)	65.9 (134.8)	53.7 (131.8)	52.9 (135.4)	50.5 (139.2)	2.52 (130.7)	2.34 (131.0)	72.12	73.50
2	1.88 (129.9)	1.65 (130.4)	1.56 (129.8)	1.61 (123.8)	5.8 (136.6)	3.64 (129.0)	1.92 (122.4)	71.8 (128.3)	75.4 (127.7)	1.62	1.48
3	4.78 (124.9)	4.94 (124.7)	4.85 (124.6)	4.61 (125.1)	5.85 (125.3)	4.83 (124.6)	4.14 (124.5)	2.73 (125.3)	2.61 (125.4)	4.70	4.67
4	1.11 (132.9)	1.18 (133.4)	1.15 (132.7)	1.00 (128.9)	1.1 (135.1)	1.09 (132.9)	0.94 (128.9)	4.1 (128.2)	4.35 (128.2)	0.76	1.0
5	(128.6)	0.54 (128.7)	0.52 (128.6)	0.27 (128.4)	0.3 (128.7)	0.34 (128.6)	0.23 (128.6)	0.2 (127.7)	0.35 (127.9)	0.71	~0
6	(126.2)	0 (126.4)	0 (126.1)	0 (126.3)	0 (126.0)	0 (126.5)	0 (126.4)	(128.3)	(128.3)	~0	~0
7	(127.6)	0.42 (127.8)	0.38 (127.4)	0 (126.9)	0.4 (129.0)	0.33 (127.9)	0 (126.7)	(126.8)	0.27 (126.8)	~0	~0

8	0.86 (125.7)	0.81 (126.2)	0.81 (126.2)	2.00 (128.1)	1.46 (125.2)	0.9 (126.5)	1.64 (125.8)	(129.3)	0 (129.4)	0.75	
9	3.53 (130.9)	3.74 (131.7)	3.65 (131.5)	2.20 (129.5)	2.5 (130.7)	2.00 (130.5)	1.77 (130.2)	4.85 (132.3)	4.99 (132.6)	3.82	2.45
10	4.01 (133.6)	4.28 (134.1)	4.21 (134.0)	3.30 (133.4)	3.20 (134.1)	3.09 (134.2)	2.84 (133.8)	0.85 (135.0)	0.88 (135.5)	4.01	3.64
CH₃		2.48 (51.6)	1.60 (28.1)	3.07, 0.2[c] (38.7, 34.7)		42.4 (29.5)	0 (27.2)		2.50 (51.5)		
Quart-C			2.69 (80.6)				39.7 (45.1)				
Carbonyl	(168.8)	(166.7)	(165.9)	(170.7)		(199.7)	(211.8)	(167.6)	(166.0)		

[a] Reference 13.
[b] Reference 6.
[c] First entry is trans, second is cis, according to a reversal of assignments (ref. 17).

TABLE 4-4. CARBON–CARBON COUPLING CONSTANTS (AND CARBON-13 CHEMICAL SHIFTS) OF INTERNALLY LABELED BENZENES AND NAPHTHALENES

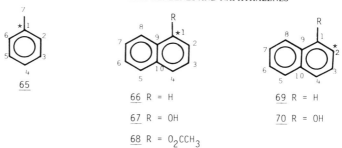

65

66 R = H

67 R = OH

68 R = O$_2$CCH$_3$

69 R = H

70 R = OH

Carbon no.	J_{CC}(Hz) to labeled carbon (δ_{C-13}, ppm) of compound:					
	65[a]	66[b]	67[b,c]	68[c]	69[d]	70[d]
1					60.3	68.84
			(151.4)	(146.9)	(128.1)	(153.1)
2	57.30	+60.4	+70.0	74.0		
			(108.8)	(118.2)	(126.0)	(108.4)
3	0.78	−2.45	<0.6	<0.6		55.96
			(125.9)	(125.4)		(126.3)
4	9.45		7.9	8.6	2.43	2.81
			(120.9)	(125.9)		(119.5)
5			+3.8	4.2	1.47	1.46
			(127.7)	(128.0)		(127.6)
6		−1.47	<0.6	<0.6		1.25
			(126.3)	(126.4)		(126.2)
7	44.15	+5.51	+4.5	5.0		0.49
			(125.3)	(126.4)		(124.6)
8			<0.6	<0.6	5.45	3.04
			(121.6)	(121.3)		(122.4)
9		+55.87	+65.4	65.5	1.69	0.50
			(124.5)	(126.9)	(133.7)	(125.1)
10		0.20	+1.8	2.0	7.97	5.77
			(134.9)	(134.8)		(135.0)

[a] Reference 18.
[b] Reference 17.
[c] Reference 19.
[d] Reference 20.

In carboxyl-labeled carboxylic acids, observed vicinal couplings are in accord with these trends and with principles established in Chapter 3 regarding aliphatic carboxylic acids. In the series **4, 5, 10, 11** it is seen that a cis coupling is small (2.38 Hz in **4**, 1.54 Hz in **11**) and that a trans coupling is much larger

TABLE 4-5. CARBON–CARBON COUPLING CONSTANTS (AND CARBON-13 CHEMICAL SHIFTS) OF INTERNALLY LABELED ANTHRACENES

71 R = H	74 R, R' = H, H
72 R = CH$_3$	75 R = =O, R' = H, H
73 R = OCH$_3$	76 R, R' = =O

J_{CC}(Hz) to labeled carbon ($\delta_{C\text{-}13}$, ppm) of compound:

Carbon no.	71[a]	72[b,c]	73[b,c]	74[a]	75[b]	76[d]
1	+1.80	0.3 (125.1)	+1.20 (122.3)	+2.35	1.69 (127.8)	+1.2
2	+5.97	+5.95 (125.6)	+5.70 (125.4)	+3.35	3.46 (127.3)	+3.65
3	−1.62	−1.72 (125.2)	−1.92 (125.6)	−0.35	0.9 (132.7)	−0.85
4	+3.13	+3.13 (129.5)	+4.20 (128.7)	+3.30	3.36 (128.8)	+3.1
9		(130.4)	(152.8)		(184.0)	
10	7.6	8.00 (125.7)	8.00 (122.4)	3.4	3.28 (32.3)	3.84
11	−0.7	0.5 (131.9)	+1.48 (132.8)	−0.2	2.83 (141.0)	+2.44
12	+60.5	(130.6)	+69.89 (124.7)	+42.0	54.90 (132.2)	+53.9

[a] Reference 17.
[b] Reference 21.
[c] Signs predicted in ref. 17.
[d] Reference 13.

(7.30 Hz in 10, 7.50 Hz in 11). It is apparent that for the cis couplings, through-space interactions via the carbonyl account for the small ^3J values, just as for aliphatic cis couplings. The trans couplings in the olefins are somewhat larger than are analogous trans couplings in aliphatics, because of the shorter C=C bond and more efficient transmission of the coupling. At first glance, it may seem surprising that the cis couplings are not also magnified because of the shorter C=C bond; however, this more compact geometry would also place the carbonyl group closer to the terminal methyl group. For 3-butenoic acid (5), the vicinal coupling (4.20 Hz) is intermediate between the cis and trans couplings of 4, 10, and 11 and is similar to the analogous vicinal coupling in the aliphatic

TABLE 4-6. CARBON–CARBON COUPLING CONSTANTS (AND CARBON-13 CHEMICAL SHIFTS) OF PHENANTHRENES

77 R = OH	**80** R = CH$_3$		
78 R = OCH$_3$	**81** R = CHO	**83**	
79 R = O$_2$CCH$_3$	**82** R = CN		

Carbon no.	J_{CC}(Hz) to labeled carbon ($\delta_{C\text{-}13}$, ppm) of compound:						
	77[a]	78[a]	79[a]	80[a]	81[a]	82[a]	83[a,b]
1							+2.70
	(154.7)	(155.5)	(146.8)	(134.3)	(131.2)	(110.6)	(128.5)
2	66.8	70.0	72.0	60.1	60.1	62.8	5.35
	(110.9)	(105.5)	(118.9)	(127.3)	(134.7)	(131.9)	(126.5)
3		0.6		1.7	1.2	1.0	~0
	(127.6)	(126.1)	(125.9)	(125.7)	(125.3)	(125.5)	(126.5)
4	8.2	7.7	8.8	8.2	7.6	9.0	+3.0
	(114.6)	(114.7)	(120.3)	(120.5)	(128.6)	(127.3)	(122.6)
5							−1.2
	(123.8)	(122.9)	(122.7)	(122.5)	(121.8)	(122.4)	
6	0.5	0.7					~0
	(127.0)	(126.4)	(126.7)	(126.0)	(127.0)	(127.6)	
7		0.4					
	(127.1)	(126.3)	(126.7)	(125.8)	(126.9)	(127.4)	
8	0.3	0.3					+5.50
	(129.1)	(128.2)	(128.4)	(128.0)	(128.3)	(128.7)	
9	4.6	4.4	5.0	5.0	4.5	5.4	+63.0
	(125.9)	(125.8)	(127.4)	(126.2)	(129.9)	(129.9)	(126.9)
10		0.7	0.6	1.5		1.2	
	(121.4)	(120.2)	(119.0)	(122.4)	(122.5)	(122.7)	
11	66.3	67.2	68.8	56.4		52.7	+53.8
	(123.3)	(122.9)	(124.6)	(130.4)	(130.0)	(130.3)	(132.0)
12	2.0	2.1	1.4				
	(132.7)	(131.2)	(131.5)	(129.9)	(130.5)	(129.4)	(130.3)
13	3.9	3.8	4.2	2.7	2.9	3.1	5.85
	(131.2)	(129.8)	(129.7)	(130.3)	(129.4)	(131.8)	(130.3)
14	1.0	0.9	1.0	1.1	0.9		−1.6
	(133.0)	(132.0)	(131.5)	(131.2)	(131.4)	(131.9)	
CH$_3$		2.2	2.0	44.6			
		(55.5)	(21.0)	(19.7)			
Carbonyl			3.7		54.8		
			(169.0)		(193.0)		
CN							(118.1)

[a] Reference 22.   [b] Reference 14.

TABLE 4-7. CARBON–CARBON COUPLING CONSTANTS (AND CARBON-13 CHEMICAL SHIFTS) OF PYRENES

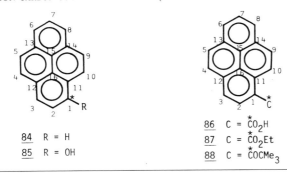

84　R = H
85　R = OH

86　C = $\overset{*}{C}O_2H$
87　C = $\overset{*}{C}O_2Et$
88　C = $\overset{*}{C}OCMe_3$

Carbon no.	J_{CC}(Hz) to labeled carbon (δ_{C-13}, ppm) of compound:				
	$84^{a,b,c}$	$85^{a,b}$	86^d	87^d	88^e
1			72.1	75.76	50.8
	(125.2)	(156.4)	(124.5)	(123.7)	(136.4)
2	+57.0	+64.5	1.75	1.72	2.2
	(126.1)	(113.7)	(128.5)	(128.5)	(122.6)
3		0.34	4.50	4.63	4.0
		(126.5)	(124.4)	(124.1)	(123.9)
4	1.55	1.67		0.4	0.2
	(127.6)	(127.9)	(127.1)	(127.2)	(127.4)
5	2.27	1.36		0.29	~0
		(124.7)	(129.3)	(129.5)	(128.3)
6		1.09, 1.39		0	~0
		(124.3)	(126.3)	(126.3)	(125.9)
7	0.76			0	~0
		(126.6)	(126.5)	(126.3)	(126.5)
8		1.39, 1.09		0	~0
		(124.7)	(126.1)	(126.1)	(125.6)
9	+5.82	+5.00		0.45	~0
	(126.3)	(126.3)	(129.0)	(129.2)	(128.3)
10	+1.55	+0.66		0.97	1.9
	(122.0)	(122.0)	(124.8)	(125.3)	(128.8)
11	+58.9	+69.0	3.52	3.63	1.7
	(131.3)		(130.2)	(131.1)	(127.7)
12	+7.71	+7.55	0.97	1.05	1.1
	(124.9)	(125.3)	(133.6)	(134.1)	(131.4)
13	0.31	0.2	0.50	0.49	0.2
	(132.4)		(130.6)	(131.0)	(131.4)
14	−1.35	−1.28		0.15	0
	(132.5)		(129.8)	(130.4)	(130.8)
15	+3.05	+3.90	0.35	0.49	
	(125.7)		(123.5)	(124.1)	(124.7)
16	0.2	$+1.5^d$	4.52	4.76	2.2
		(126.7)	(124.1)	(124.7)	(123.8)

[a]Reference 20. [b]Signs predicted in ref. 17. [c]Carbons 13 and 14 reversed from original listings (ref. 17). [d]Reference 23. [e]Reference 13.

TABLE 4-8. CARBON-CARBON COUPLING CONSTANTS (AND CARBON-13 CHEMICAL SHIFTS) OF PHENYL-SUBSTITUTED ETHANES, ETHYLENES, AND ACETYLENE

Structures (labeled carbon marked with *):

- **89** Ph—*CH₂—CH₂—Ph (ring carbons numbered 1–10, X substituents at 6 and 10)
- **90** Ph—CH(OH)—*CH(OH)—Ph
- **91** Ph—CH—(O)—*CH—Ph (epoxide)
- **92** Ph—CH(Ph)—*CH(Ph)—Ph
- **93** Ph—*C(OH)(Ph)—C(OH)(Ph)—Ph
- **94** Ph—*C(Ph)—(O)—C(Ph)—Ph
- **95** Ph—CH=*C(H)—Ph
- **96** Ph—*C(H)=C(H)—Ph
- **97** Ph₂C=*CPh₂
- **98** Br—Ph—*C(Ph)(Br)=C(Ph)(Br)—Ph—Br
- **99** Ph—C(=O)—*C(=O)—Ph
- **100** Ph—*C(=N—NH₂)—C(=N—NH₂)—Ph (H₂N)
- **101** Ph—*C≡C—Ph

J_{CC} (Hz) to labeled carbon ($\delta_{C\text{-}13}$, ppm) of carbon no.:

Compound no.	1	2	3	4	5	6	7	8	9	10	References
89		+32.6 (38.1)	+43.5 (141.7)	+2.5 (128.5)	3.2 (128.4)	0.8 (126.0)		+2.2	1.1	0.8	24
90		(79.1)	(139.9)	+2.8 (127.0)	3.15 (128.1)	0.6 (127.9)	−1.9	+1.3	0.8	0.6	24
91		+28.1 (58.8)	+56.6 (137.2)	+3.8 (125.5)	4.15 (128.6)	0.8 (128.3)	−1.1	+1.5	0.6	0.8	24
92		+33.8 (56.4)	+43.1 (143.5)	+2.2 (128.5)	3.3 (128.1)	0.75 (125.8)	−0.6	+2.0	0.55	0.75	24
93		+41.3 (83.1)	+48.5 (144.3)	+1.8 (128.7)	3.2 (127.3)	1.0 (126.9)	+0.3	+0.9	0.7	1.0	24
94		+28.1	+57.8	+2.5	3.9	1.1	−0.5	+1.25	0.6	1.1	24

95	(73.9) +72.0	(138.7) +57.0	(128.3) +2.2	(127.6)	(127.1) −1.2	+0.1	+5.0		+0.6	14, 24
96	(128.6)	(137.3)	(126.5) +2.1	(128.6) 3.8	(127.5) 0.75		+3.6	0.5		14
97	(130.3) +76.9	(137.2) +54.4	(128.9) +2.64	(128.2) 3.66	(127.1) <0.7	+1.0	+1.44			14
98	(141.0) +77.4	(143.8) +54.7	(131.3) +2.3	(127.7) 3.85	(126.4) −0.7	+1.1	+2.0	0.85	+1.3	14
99	(139.5) 51.8	(141.3) +54.5	(132.6) +3.4	(131.2) 4.37	(121.1) 1.02	+14.0	+0.4	0	0	9
100	48.8	+66.0	+1.8	4.47	1.0	+4.9	+1.2	0	0	9
101	185.0	+91.1	+2.62	5.56	1.0	+13.1	+1.76	0	0.75	9

TABLE 4-9. CARBON–CARBON COUPLING CONSTANTS (AND CARBON-13 CHEMICAL SHIFTS) OF PHENYL-SUBSTITUTED CYCLIC COMPOUNDS[a]

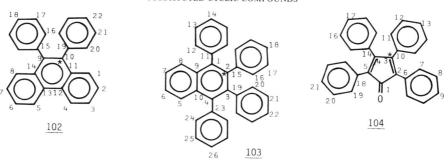

| Carbon no. | J_{CC}(Hz) to labeled carbon ($\delta_{C\text{-}13}$, ppm) of compound: | | |
	102	**103**	**104**
1	+1.84		9.07
	(127.8)	(138.5)	(200.2)
2	4.33		+74.76
	(126.5)	(138.9)	(125.4)
3		+55.28	
	(126.3)		(154.6)
4			+48.0
	(122.4)		
5		−1.26	0.40
		(127.0)	
6			+0.94
		(125.3)	(130.8)
7	0.6	1.26	3.28
			(130.2)
8	+4.20	+4.08	
			(128.0)
9	+67.2	0.34	
	(136.8)	(132.1)	(127.5)
10		5.97	+57.78
			(133.2)
11	+54.3	+1.3	2.14
	(131.8)	(140.5)	(129.4)
12	0.4		3.78
	(129.9)	(131.3)	(128.0)
13	5.24	0.84	
		(127.5)	(128.5)
14	0.1	1.12	2.08
		(126.4)	
15	+1.3	+56.31	0.88
	(139.3)	(139.6)	

TABLE 4-9. (continued)

Carbon no.	J_{CC}(Hz) to labeled carbon (δ_{C-13}, ppm) of compound:		
	102	103	104
16	+1.1	~1.9	
	(130.8)	(131.3)	
17	0.6	3.47	
	(127.5)	(126.6)	
18		1.58	+4.79
	(126.3)	(125.9)	
19	+56.2	+1.80	−0.89
20	+2.1		
21	3.42	0.95	
22			
23		+3.4	

[a]Reproduced with permission.[14]

butanoic acid (3.6 Hz, Chapter 3). It is perhaps not surprising that butanoic acid and 3-butenoic acid exhibit similar vicinal couplings, for they are both conformationally mobile and possess comparable geometries. The ^2J values for the geminal couplings in **4, 5, 10**, and **11**, are near zero for the cis olefin **4** and larger for the trans olefin **10**. These ^2J values are mirrored by $^2J = 0.30$ Hz in *cis*-cinnamic acid and $^2J = 1.66$ Hz in *trans*-cinnamic acid[6] (**8** and **9**, Table 4-1).

It was noted in Chapter 3 that attempts to generate a Karplus-like plot for olefins of the general structure C=C—C—C fail, because ^3J vs ϕ simply generates a horizontal plot of data points (Figure 3-4). Accordingly, it is not surprising that ^3J values of 1,4-dihydrobenzoic acid (**13**) and of 1,4-dihydro-1-naphthoic acid (**14**) are similar (3.2 and 3.3 Hz, respectively, Table 4-1), even though **13** is flat and **14** is puckered.[7] The practicing organic chemist should be mindful of this phenomenon in conformational analysis. However, there appear to be situations where the value of ^3J can reflect the degree of torque about C—C—C—C linkages in olefins,[14] although in order to explore these systems we must briefly cross into the domain of totally conjugated systems. In the series **95–97**[14,24] (taken from Table 4-8) the ^3J value drops progressively from 5.0 Hz for *trans*-stilbene (**95**) to 3.6 Hz for *cis*-stilbene (**96**) to 1.44 Hz for

95   96   97

TABLE 4-10. CARBON–CARBON COUPLING CONSTANTS (AND CARBON-13 CHEMICAL SHIFTS) OF FLUORENE AND ACENAPHTHENONE COMPOUNDS[a]

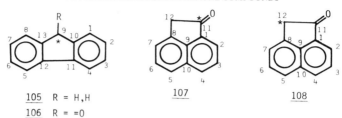

105   R = H,H
106   R = =O

107

108

Carbon no.	J_{CC}(Hz) to labeled carbon (δ_{C-13}, ppm) of compound:			
	105	**106**	**107**	**108**
1	3.12 (125.8)	2.62 (124.4)	54.81 (134.9)	17.58
2	3.75 (127.5)	3.75 (129.8)	1.83 (120.8)	0
3	0.50 (127.5)	0.88 (135.4)	4.03 (127.4)	0
4	2.50 (120.6)	3.75 (121.4)	0.98 (130.4)	0
5			0 (123.4)	0
6			0 (127.8)	3.52
7			4.15 (120.5)	0
8			3.66 (133.9)	45.12
9	(37.3)	(205.7)	9.80 (142.2)	1.56
10	42.25 (144.0)	(133.0)	<2.0 (130.3)	1.17
11	4.50 (145.1)	7.88 (145.1)	(190.4)	
12			42.72 (42.0)	

[a]Reproduced with permission from Wiley Heyden Ltd.[2]

tetraphenylethylene (**97**). Increased twist of the phenyl ring is clearly the responsible factor here,[14] leading to values of ϕ closer to 90°. It appears that the dihedral angular relationship works in these cases because the coupling pathway is fully conjugated. Therefore, when the phenyl substituent approaches planarity with the conjugated olefin an efficient coupling mechanism is available. Further,

TABLE 4-11. CARBON–CARBON COUPLING CONSTANTS OF TRICYCLIC NAPHTHALENE COMPOUNDS[a]

| Carbon no. | J_{CC}(Hz) to labeled carbon of compound: | | | | |
	109	**110**	**111**	**112**	**113**
1	51.64	53.56	51.55	76.40	60.8
2	1.21	10.99	2.74	3.73	3.44
3	4.06	4.56	3.77	5.23	4.57
4	1.10	0.89	0.93	1.10	0.86
5	0.65	0.40	~0	~0	~0
6	~0	~0	~0	0.39	~0
7	0.52	0.21	~0	3.27	3.75
8	2.9	0.65	0.55	3.57	6.57
9	3.05	2.85	1.72	6.55	6.87
10	3.62	2.58	3.00	4.12	3.62
11					
12	41.02	39.24			
13	2.13	2.65			

[a]Reproduced with permission.[6]

when ϕ approaches 90°, apparently the hyperconjugative contributions are not as effective in totally sp^2 linkages. Example **104**[14] bears out this point; the torsional angles in **104** can be approximated by the known X-ray parameters of

1,2,3,4-tetraphenylcyclopentadiene.[27] The 1-phenyl group is twisted to ca. 30° and exhibits a ^3J value of 3.28 Hz. The 2-phenyl group is twisted more (ca. 70°) and shows a much smaller ^3J value of 0.88 Hz.

In acetylenic systems, we have a class of compounds whose coupling behavior

TABLE 4-12. CARBON–CARBON COUPLING CONSTANTS (AND CARBON-13 CHEMICAL SHIFTS) OF METHYLENE BENZOCYCLOALKENES[a]

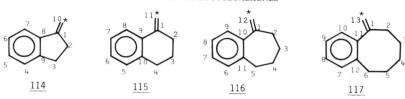

Carbon no.	J_{CC}(Hz) to labeled carbon (δ_{C-13}, ppm) of compound:			
	114	**115**	**116**	**117**
1	73.9 (150.5)	72.6 (143.4)	71.0 (153.8)	71.1 (152.9)
2	<0.6 (31.2)	1.5 (33.4)	2.0 (36.7)	2.2 (41.9)
3	2.1 (30.0)	2.8 (24.0)	1.9 (31.6)	2.9 (27.0)
4	(125.3)[b]	(30.6)	(27.5)	(26.4)
5	(126.4)[b]	(125.8)[b]	(36.5)	(31.9)[b]
6	(128.2)[b]	(127.2)[b]	(126.8)[b]	(32.2)[b]
7	1.7 (120.6)	(129.2)[b]	(127.8)[b]	(125.8)[b]
8	(141.1)	4.2 (124.3)	(129.7)[b]	(127.1)[b]
9	4.8 (146.7)	(134.9)	2.4 (128.9)	(128.9)[b]
10	(102.4)	3.6 (137.1)	(128.9)	(127.5)
11		(107.8)	2.4 (141.0)	2.2 (143.8)
12			(114.3)	2.2 (140.0)
13				(113.8)

[a]Reproduced with permission.[25]
[b]Assignments indefinite.

is quite different. Geminal couplings in acetylenes transmitted through the triple bond are quite large: in **15–18** the ^2J values lie in the range 11.8–20.3 Hz. In contrast, vicinal couplings are surprisingly small. With $^3J = 1.76$–1.84 Hz in **16** and **17**, it is seen that the shorter triple bond and increased s character of the associated bonds do not produce a very large vicinal coupling. Indeed, these vicinal couplings are smaller than for average aliphatic vicinal couplings. It is manifest that whereas olefinic couplings find many parallels with aliphatic couplings, acetylene couplings are unique.

$$\underset{\textbf{15}}{\overset{1.76}{CH_3-C\equiv \overset{*}{C}-H}} \qquad \underset{\textbf{16}}{\overset{13.1}{C\equiv \overset{*}{C}}} \qquad \underset{\textbf{17}}{\overset{1.84\quad 20.3}{CH_3-C\equiv C-\overset{*}{C}O_2CH_3}} \qquad \underset{\textbf{18}}{\overset{18.42}{CH_3O_2C-C\equiv C-\overset{*}{C}O_2CH_3}}$$

α-Labeled Benzene Compounds

As noted in the previous section, couplings in the first category of π systems also include systems such as **55** and **87** where the labeled carbon is a substituent, even though the entire coupling route is conjugated. The reason for the inclusion for such systems is that their coupling behavior is similar to analogous systems where the labeled carbon is sp³ hybridized, such as toluene (**19**). Accordingly, in various α-labeled benzene derivatives (Table 4-2) the vicinal couplings in **19–43** fall in the range of 3–6 Hz, the geminal couplings fall in the range of 1.5–4 Hz, and the four-bonded couplings are less than 1.6 Hz. This consistent behavior of the carbon couplings persists whether the hybridization of the α-labeled carbon is sp³, sp², or sp. Instead, overriding considerations for the values of vicinal couplings appear to be both the hybridization of the substituent and the presence of electronegative substituents on the α carbon, whereas the behavior of ^2J and ^4J is more subtle. Examples below illustrate these features. Vicinal

19
−3.10
−3.84
0.86

21
*CH₂Cl
−3.69
−4.23
0.69

43
*CCl₂ / Ph
−3.21
−4.52
−1.02

25
O=*C−H
−3.78
−4.61
1.14

27
O=*C−Cl
−3.53
−5.46
−1.18

29
*CN
−2.61
−5.75
−1.59

101
*C≡C−Ph
−2.62
−5.56
−1.0

couplings for the sp³ systems (**19**, **21**, and **43**) are ca. 4.2 Hz; for sp² systems (**25** and **27**) they are larger (ca. 5 Hz); and for the sp systems (**29** and **101**) they are a little larger yet (ca. 5.6 Hz). Such dependence of ^3J on the hybridization of a terminal substituent is not surprising, because the efficiency of coupling increases as the s character of associated bonds increases. This behavior was

TABLE 4-13. CARBON–CARBON COUPLING CONSTANTS (AND CARBON-13 CHEMICAL SHIFTS) FOR BENZANTHRACENE AND BENZOPYRENE DERIVATIVES[a]

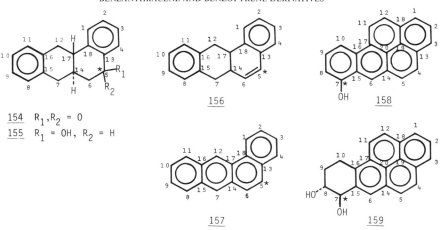

Carbon no.	J_{CC}(Hz) to labeled carbon (δ_{C-13}, ppm) of compound:					
	154	**155**	**156**	**157**	**158**	**159**
1	3.5	2.9	3.1	2.9		
	(126.5)	(125.8)	(123.8)	(122.7)	(125.0)	(123.2)
2	(133.6)	(126.7)	(127.1)	(126.5)	(126.7)	(126.2)
3	5.0	3.3	4.7			
	(126.1)	(125.6)	(126.2)	(126.5)	(125.7)	(125.6)
4			2.5	2.6		
	(126.9)	(126.7)	(125.9)	(128.3)	(126.9)	(128.0)
5	(197.0)	(68.1)	(127.3)	(126.8)	(127.8)	(128.3)
6	40.8	36.9	67.4		<0.5	2.1
	(45.3)	(39.8)	(132.5)	(126.8)	(120.1)	(125.4)
7	5.1	5.3	3.5			
	(36.8)	(37.9)	(36.1)	(127.1)	(155.0)	(76.6)
8					69.3	44.1
	(128.9)	(128.9)	(129.0)	(128.2)	(108.5)	(74.1)
9					<0.5	<0.5
	(126.0)	(125.4)	(125.6)	(125.4)	(129.3)	(135.1)
10					7.9	3.6
	(126.0)	(125.4)	(125.6)	(125.5)	(114.7)	(122.9)
11	(128.4)	(128.4)	(128.6)	(127.5)	(123.3)	(122.6)
12				1.1		
	(34.8)	(37.1)	(33.3)	(121.3)	(127.8)	(127.6)
13	51.0	45.4	52.4	65.0		
	(132.1)	(140.7)	(133.7)	(131.7)	(132.6)	(131.7)
14			0.6		4.5	3.2
	(36.4)	(36.9)	(35.7)	(131.7)	(129.4)	(131.3)

TABLE 4-13. (continued)

Carbon no.	J_{CC}(Hz) to labeled carbon (δ_{C-13}, ppm) of compound:					
	154	**155**	**156**	**157**	**158**	**159**
15					64.8	46.2
	(134.8)	(135.8)	(135.4)	(131.7)	(124.0)	(137.6)
16					1.9	2.1
	(134.8)	(135.8)	(135.3)	(128.6)	(130.6)	(127.4)
17	1.6		4.3	6.9	3.9	3.1
	(38.9)	(40.2)	(37.9)	(130.3)	(127.8)	(126.8)
18	6.3	2.4		1.3		
	(146.0)	(138.6)	(137.5)	(130.4)	(132.4)	(132.6)
19					(126.3)	(128.3)
20					1.0	<0.5
					(124.6)	(125.0)

noted previously in aliphatic systems (Chapter 3) for transoid couplings. The dependence of ^3J on electronegative substituents is also exemplified by the same compounds. Thus, ^3J increases through the sp^3 series **19, 21, 43** as more chloro groups are substituted on the α carbon (3.84–4.52 Hz); and the same trend is noted for the sp^2 compounds **25** and **27** (4.61–5.46 Hz) as a chloro group is substituted. Again, it was noted previously for aliphatic systems (Chapter 3) that an electronegative terminal substituent increases the value of ^3J, and accordingly analogies between aliphatic systems and π systems continue. In geminal and four-bonded couplings in the same examples, no discernible trends are immediately apparent. Instead, for these couplings other factors must be considered (see the following section).

Internal substituents also can play an important role in determining the value of vicinal couplings. If the analogy between aliphatic and π systems applies here also, then these vicinal couplings should be attenuated. Examples **44–47** show that this is indeed the case:[15] vicinal couplings to C-3 (on the same side as the

substituent) are ca. 1.5–2.4 Hz, substantially smaller than vicinal couplings to C-5 (away from the substituent) which are just about the same value as that in toluene (**19**, 3.84 Hz). Further, these attenuated vicinal couplings are smaller in those compounds whose substituents have a greater electronegativity (**44** and **45**, $^3J \approx 1.6$ Hz). The presence of an internal alkyl substituent has little effect on vicinal couplings: in compound **40**[3] the two vicinal couplings differ by less than 0.5 Hz and in other systems the difference can be even less. Geminal couplings also are attenuated by these electronegative substituents: these couplings to C-2 in **44–47** are 1 Hz or less. Four-bonded couplings, in contrast, are about the same as before, probably because there exists a coupling path on the side away from the substituent.

Closer scrutiny of compounds **44–47** and **40**, above, discloses an apparent anomaly. Increasing the electronegativity of an internal substituent reduces the values of ^2J and ^3J, and because an alkyl group has an electronegativity similar to hydrogen it is not surprising that the methyl group in **40** does not much affect the vicinal couplings. However, there is a substantial difference between the two geminal couplings in **40** (2.54 and 3.22 Hz), and other 2-methyl compounds in Table 4-2 exhibit irregular behavior. For example, in *o*-toluic acid (**32**) the geminal coupling to C-2 is larger than that to C-6 (3.20, 2.38 Hz), whereas in

2-methylacetophenone (**35**) the reverse is seen[3, 13] (1.86, 3.57 Hz). These observations can be generalized for carboxyl compounds such as **32** and ketones such as **35**,[13] and it has been proposed[13] that an orientation effect of the carbonyl group is being experienced. However, it is not clear why the same orientation of the carbonyl group in **32** and **35** results in contrasting behavior. Proposed reasons for the variations in ^2J include twisting from planarity of the carbonyl group with the aromatic ring (reducing the extent of conjugation)[13] and through-space contributions,[3] but it is not understood why carboxyl derivatives behave in opposite fashion from ketones. Other variations in structure are seen in **38** and **34**. In the hindered **38**, the pivaloyl group rotates and reduced geminal couplings are observed. In the aldehyde **34**, where a preferred orientation of the carbonyl does not exist,[28] the two geminal couplings are more nearly equal. Compounds **40** and **41** are sp^3 derivatives which likewise show a large variability of ^2J. In these compounds $^2J_{C-2}$ increases and $^2J_{C-6}$ decreases when moving from **40** to **41** as the electronegative hydroxyl group moves more toward the 2-CH$_3$ group. More light will be shed on the subject of geminal aromatic couplings in the next section.

Some variability also exists in the vicinal couplings of the carbonyl compounds **32**, **35**, **34**, and **38**, but the range of ^3J is less than of ^2J. Of interest are the ^3J values in **38**, which are substantially smaller than those in **32**, **35**, or **34**. The pivaloyl group in **38** is expected to be twisted out of planarity with the aromatic ring[13] but calculations suggest[13] this twisting should have little effect on the transoid coupling to C-3 or to C-5. Further, in compounds **36** and **37**, where no steric hindrance exists between the pivaloyl group and the methyl group, the vicinal couplings are still small as in **38**. A search for the reasons for the reduced value of ^3J in **36–38** naturally leads to suggestions of distortions about the pivaloyl group, the nature of which is uncertain.

In summary, it appears that the orientation of a substituent on the α carbon of benzene derivatives may influence to a significant degree the magnitudes of geminal and vicinal couplings to aromatic carbons. However, these types of effects are not completely understood at present.

In contrast to the difficulties discussed immediately above, the role of the orientation of an α substituent in determining the magnitudes of vicinal couplings to a 2–CH$_3$ carbon is better understood. This understanding is founded on principles discussed in Chapter 3 for aliphatic couplings: for cis couplings, a terminal electronegative substituent increases ^3J as it becomes oriented more nearly away from the vicinal carbon. Compounds **32–35**, and **38–42** also show the same kind of behavior, for both sp^2 and sp^3 systems.[3, 13] In **32–35** the vicinal couplings are small when a carbonyl group is directed toward the 2–CH$_3$ groups (0.7, 0.75, and 0.91 Hz, respectively, in **32**, **33**, and **35**). In the aldehyde **34** a larger value of 1.88 Hz is seen. Because this compound is a 50:50 mixture of cisoid:transoid conformers,[28] the "real" vicinal coupling in the conformation shown in **34** should be larger, perhaps ca. 3 Hz. In the pivaloyl derivative **38** an intermediate value of ^3J is seen (1.33 Hz), consistent with the idea that the carbonyl group is (on the average) in an orientation intermediate between that in **35** and **34**, twisted away from planarity with the aromatic ring.

Calculations have been performed[3] which indicate that, as for aliphatic systems (Chapter 3), this attenuation of ^3J in **32**, **33**, and **35** is caused by through-space interactions. Turning to sp³-labeled compounds (**39–41**), a similar attenuation is found in ^3J as the substituent is rotated more nearly toward the coupled carbon (compare 3.15 Hz in **40** with 2.82 Hz in **41**). The magnitude of this change is on the same order as for aliphatic primary and secondary alcohol cis couplings (see compounds **50** and **53** in Chapter 3). Finally, the sp system **42** shows a ^3J value of 1.95 Hz which is small, in view of the increased s character of the α substituent. That this small value of ^3J is a result of negative through-space contributions between the nitrile group and the methyl group has been indicated again by calculations[3] wherein orbital cutoffs between the coupled carbons have been performed.

Geminal Couplings in Polynuclear Compounds

Compounds **54, 61, 124,** and **86** summarize the couplings in α-labeled compounds. In this series carboxylic acids have been chosen to represent trends because more data exist for carboxylic acids than for any other functionality. Benzoic acid (**22**) and *o*-toluic acid (**32**) are also included to complete the series and the comparison. The geminal coupling in **22** of 2.54 Hz is roughly approximated in the geminal couplings in the other carboxylic acids. Variations principally occur at the quaternary position: the ^2J values are generally larger to the quaternary positions of **54** (3.53 Hz), **86** (3.52 Hz), and **32** (3.20 Hz) and thereby suggest a substituent effect. However, the quaternary ^2J value in 9-anthroic acid (**124**) is much smaller (1.9 Hz). Hence, it seems that there is an orientation effect of the carbonyl group.[13] In **54**, **86**, and **32** the preferential direction of the C=O group toward the quaternary carbon appears to attenuate these ^2J values. In **124**, of course, there are two equivalent orientations, and in any case two peri interactions would disallow as great a degree of planarity of the —CO_2H group with the aromatic hydrocarbon. The conclusion is that orientations of a carbonyl toward an aromatic carbon atom in a carboxylic acid increase the coupling (above 3 Hz), or orientations away from an aromatic carbon atom produce a smaller ^2J value (2–3 Hz), whereas orthogonal orientations result in a smaller coupling yet (<2 Hz). An interesting test compound for these generalizations is afforded by biphenyl-2-carboxylic acid.[2] In the expected preferred orientation for this compound (**122**), a larger coupling is expected to C-1 than to C-3, and such behavior is indeed seen (2.68 and 1.70 Hz, respectively).

These conclusions regarding carboxyl compounds may be extended to ketones. However, it must be remembered from the discussion regarding

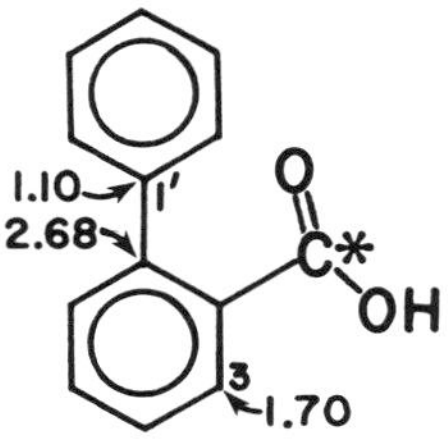

122

benzene compounds that with respect to geminal couplings, ketones behave in opposite fashion from carboxylic acids, i.e., the relative magnitudes of the ^2J values are reversed. Inspection of sample compounds below bears out these expectations. In these compounds (59, 60, 88) the effect of a bulky group is demonstrated. In unhindered 59 the ^2J value to C-9 (2.00 Hz) is smaller than the

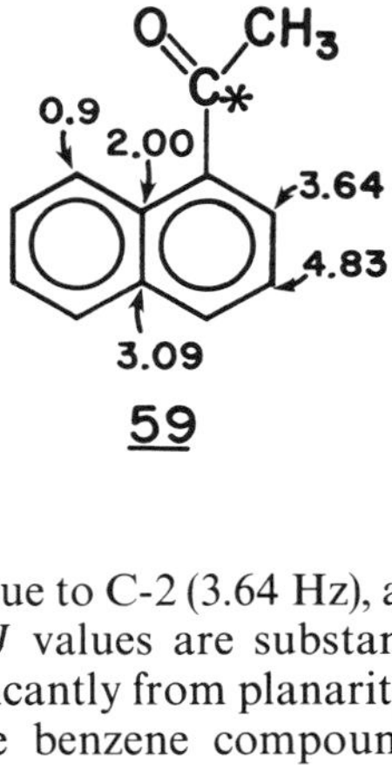

59 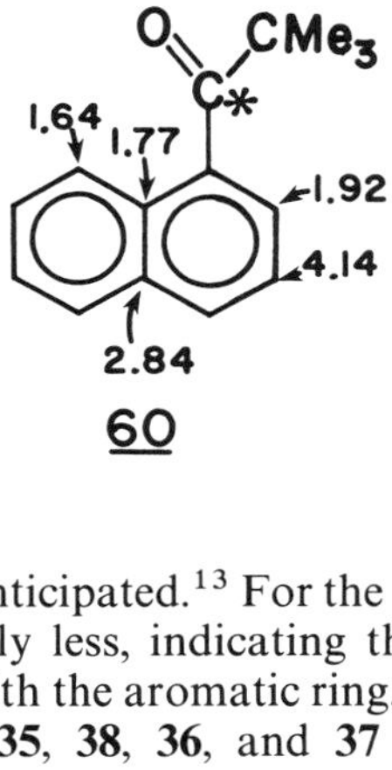   **60**

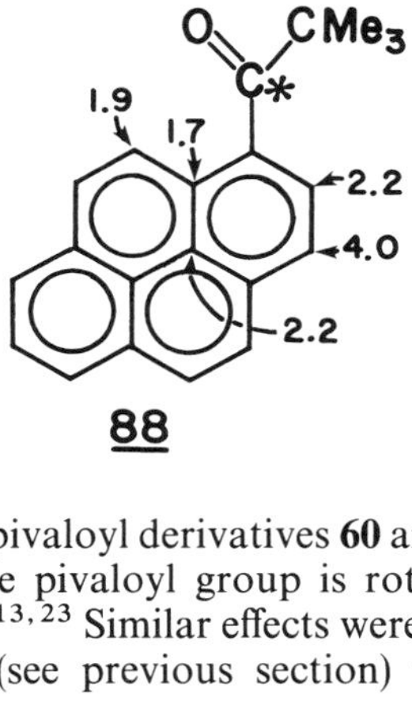

88

^2J value to C-2 (3.64 Hz), as anticipated.[13] For the pivaloyl derivatives 60 and 88 the ^2J values are substantially less, indicating the pivaloyl group is rotating significantly from planarity with the aromatic ring.[13, 23] Similar effects were seen in the benzene compounds 35, 38, 36, and 37 (see previous section) using

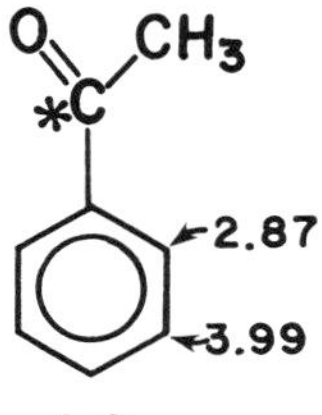

26

acetophenone (26) as a standard.[3, 13] In 26 the "normal" geminal coupling of 2.87 Hz is observed. In 2-methylacetophenone (35) a smaller ^2J (1.86 Hz) is noted at the quaternary carbon, whereas a larger ^2J (3.57 Hz) is seen at the unhindered 6 position, values quite consistent with the respective values for the naphthalene compound **59** (2.00 and 3.64 Hz). In **38** with the bulky pivaloyl group, the ^2J values are reduced (1.47 and 2.30 Hz), in agreement with the observations above for **60** and **88**. Even in 4-methylpivalophenone (**36**) and 3-methylpivalophenone (**37**) ^2J values are observed to be less than those for acetophenone (**26**); the pivaloyl group is apparently twisting enough to reduce the geminal values. It is also interesting to compare these data with the ^2J values of (at first sight) a quite different system, [2.2]paracyclophane (**126**).[30] The ^2J values of **126** (1.7 and 3.9 Hz) are quite similar to those of 2-methylacetophenone (**35**) (1.86 and 3.57 Hz) and indicate that **35** is a good model for the paracyclophane compound.

126

Two compounds that amply demonstrate the above conclusions regarding geminal couplings in ketones are the cyclic compounds **109** and **111**.[6] It was concluded above that geminal behavior was adequately explained by assuming that ^2J values to carbons cisoid to the ketonic carbonyl group are small and ^2J values to carbons transoid to the carbonyl group are large. In **109** where the carbonyl is cisoid to C-2 and transoid to C-8, the respective couplings are 1.21 and 3.05 Hz. In **111** the carbonyl is cisoid to C-9 and transoid to C-2—the respective couplings are 1.72 and 2.74 Hz. In compound **110**[6] the carbonyl is cisoid to C-9 and the ^2J value is 2.85 Hz; here the carbonyl is apparently projected farther away from C-9 and the contribution via the carbonyl is not

109   **111**   **110**

nearly as effective and appears "normal." Another compound is fluorenone (**106**)[2] whose geminal geometry is the same as in **110**, and indeed the coupling (2.62 Hz) is about the same. In **107**[2] the geminal geometry is like that in **109**, except that the carbonyl is projected more away from the C-2 geminal carbon. As anticipated, the effect of the carbonyl is not as great and the ^2J value is

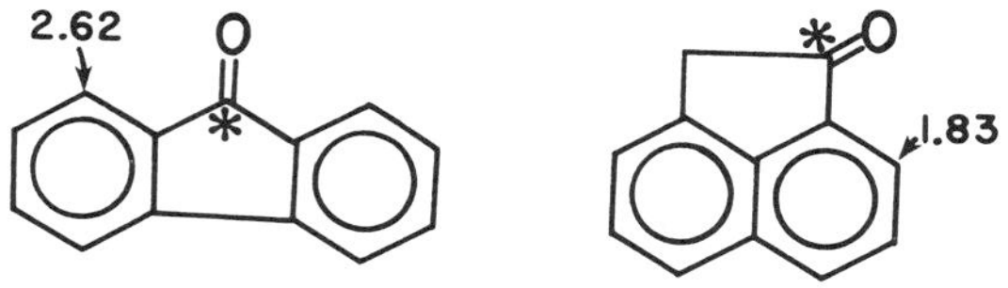

106   **107**

somewhat larger (1.83 Hz). (In compounds **110**, **106**, and **107** the other geminal couplings are not portrayed here, because these are actually multiple-path couplings and are discussed in Chapter 5.)

The behavior of aldehydes is similar to that of ketones, with the minor difference that ^2J values in aldehydes are larger. For example, in benzaldehyde (**25**) the ^2J value is 3.98 Hz (compared with 2.87 Hz in acetophenone). In l-naphthaldehyde (**58**)[13] the coupling to C-2 is larger than to C-9 (5.8 and 2.5 Hz,

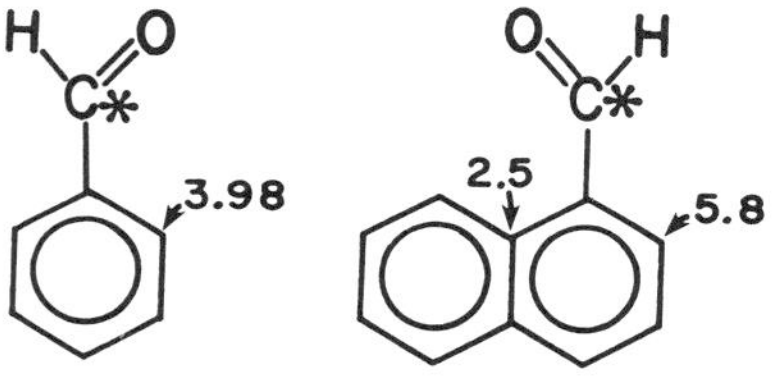

25   **58**

respectively). This observation suggests, from our conclusions regarding ketones, that the carbonyl is preferentially directed toward the C-9 carbon, and such is the case.[13]

Finally, we can briefly consider sp³-substituted polynuclear compounds, although the scanty data available permit few generalizations. In each of the systems **119** and **120**,[2] different ^2J values are observed to the two geminal

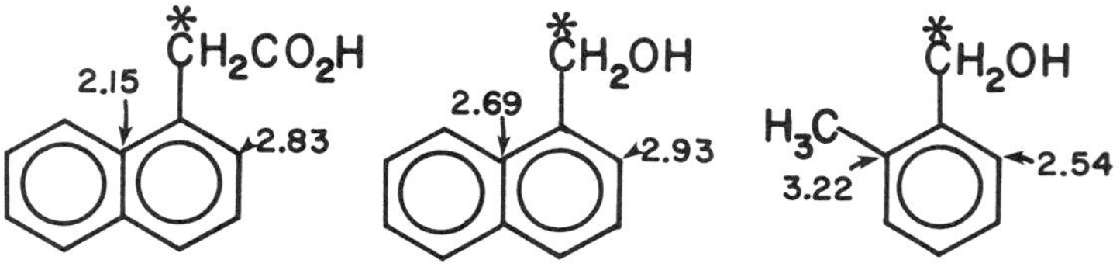

119   **120**   **40**

positions, which may indicate an orientation effect of the substituent. It is perplexing that the smaller ^2J value in **120** is the quaternary position, whereas in 2-methylbenzyl alcohol (**40**) the coupling to the quaternary position is larger. It is clear that modeling sp³-labeled polynuclear compounds upon 2-methyl-benzene compounds will not proceed as for analogous sp² systems discussed above.

Vicinal Couplings in Polynuclear Compounds

The reader is directed again to the carboxylic acids listed on p. 159. The vicinal couplings in this assemblage may be divided into two different types, cisoid or transoid. Cisoid couplings (127) are found between substituents and peri carbons and transoid couplings are between substituents and carbons in the same ring, such as 128 or 129. Cisoid couplings in these carboxylic acids are

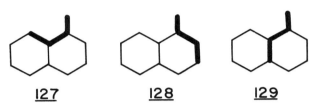

127   **128**   **129**

generally quite small—in 1-naphthoic acid and in pyrene-1-carboxylic acid (54 and 86) the ^3J (cisoid) values are less than 1 Hz. A ready explanation can be couched in terms of through-space negative contributions via the carbonyl group (see Chapter 3). In 9-anthroic acid (124) a larger ^3J value of 2.4 Hz is observed; in this compound two average conformations are possible, "diluting" the effect to each peri carbon. Additionally, steric twisting of the —CO_2H group in 124 would lead to a larger value of ^3J, because the carbonyl group would be further from the peri carbons. It should be recalled that analysis of the geminal couplings in 124 in fact strongly indicates such twisting.

Cisoid couplings for ketones, also, exhibit similar dependence on twisting of the substituent. Inspection of the compounds 59, 60, and 88 (in the previous section) again reveals a small ^3J value to the peri carbon (0.9 Hz) in the unhindered acetylnaphthalene (59). In the hindered pivaloyl derivative (60) the same cisoid coupling is larger (1.64 Hz). Likewise, in l-pivaloylpyrene (88) the cisoid coupling is 1.9 Hz. Again, it is to be recalled that analysis of the geminal couplings in these three compounds indicated twisting of the pivaloyl group in 60 and 88.

In aldehydes the same phenomenon is observed. The coupling to the peri carbon in 1-naphthaldehyde (58) is 1.46 Hz while the carbonyl group is directed toward the peri carbon.[13] This value of ^3J is larger than the one for 1-naphthoic acid (0.86 Hz) and for 1-acetylnaphthalene (0.9 Hz) but is understandable in

25   **58**

view of larger ^3J values of aldehydes vs. acids or ketones [^3J in benzaldehyde (**25**) is 4.61 vs. 4.53 Hz in benzoic acid (**22**) and 3.99 Hz in acetophenone (**26**)].

This cisoid interaction is also observed in biphenyl-2-carboxylic acid (**122** in the previous section). Just as the geminal couplings indicated a preferred orientation of the C=O group toward the phenyl substituent, so does the vicinal coupling to the 7-carbon. This small ^3J value of 1.10 Hz is quite in accord with the idea of the carbonyl group being directed toward the quaternary carbon of the phenyl substituent.

In all of the examples above, the substituent is of course conformationally mobile, and an average of the total conformational populations is actually reflected in the observed ^3J (cisoid) value. It would be instructive to see ^3J (cisoid) values in systems where the orientation of the carbonyl group is fixed. Such values are afforded by compounds **110** and **111**.[6] The indicated vicinal couplings of ca. 0.6 Hz demonstrate the effectiveness of the carbonyl group to

attenuate couplings if held rigidly in this proper orientation. These are indeed remarkably small values for planar vicinal couplings!

Finally, ^3J (cisoid) values of the sp^3-labeled systems **119** and **120**[2] are listed below. The paucity of data for sp^3 systems allows little to be concluded, except

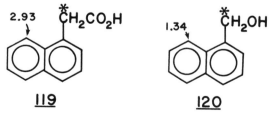

that ^3J (cisoid) couplings can be larger than those in the analogous carbonyl compounds, perhaps attenuating a nagging suspicion that all vicinal couplings to peri carbons are small.

In regard to transoid vicinal couplings (**128** and **129**), the reader's attention is directed once again to the list of carboxylic acids on p. 159. These vicinal couplings fall in the range of 4.0–4.9 Hz and therefore appear to behave more consistently than the more variable geminal couplings. The variation that does occur appears to depend roughly on the π-bond order of the coupling path. Thus, if the π-bond order is high for the center bond as in **130**, the coupling is larger. Structures **131–136** illustrate this principle. In these structures ^3J (transoid) values are listed; beside each J value is listed the Hückel molecular orbital (HMO) π-bond order (p value) for the center bond, taken from the parent hydrocarbon.[29] It is to be recalled that for crotonic acid (*trans*-2-butenoic acid, **10**, Table 4-1), whose p = 1.00 for the center bond of the associated hydrocarbon, the very large ^3J value of 7.30 Hz is observed. For benzoic acid **131** the "normal aromatic" value of 4.53 Hz is observed. In 1-naphthoic acid the coupling to the 10 position is smaller (4.01 Hz) and to the 3 position is larger (4.78 Hz), and it is seen from the Kekulé formalism (**132**) or from the corresponding p values that the larger coupling does have a higher π-bond order for the center bond. For 2-naphthoic acid (**133**) the same behavior is again observed—^3J = 4.85 Hz to C-9 and ^3J = 4.1 Hz to C-4, with the larger coupling associated with a higher π-bond

order of the center bond. For 9-anthroic acid (**134**) and pyrene-1-carboxylic acid (**135, 136**) the dependence of ^3J upon the p value is not so rigorous, but the overall trend still persists; in these compounds both the J values and the π-bond orders are about the same as, or to the low side of, the values in the standard benzoic acid (**131**). These trends in 1-naphthyl and 2-naphthyl derivatives never diverge through a large series of substituents: $-CO_2H$, $-CO_2Me$, $-CO_2CMe_3$, $-CONMe_2$, $-CHO$, $-COMe$, $-COCMe_3$, $-CH_2OH$, and $-CH_2CO_2H$ (see **54–64**, Table 4-3; **119** and **120**, Table 4-14). The sole exception to this generalization about the dependence of ^3J on π-bond order is observed in 1-pivaloylpyrene (**88**),[13] whose two transoid couplings are now quite different. It

TABLE **4-14.** CARBON CARBON COUPLING CONSTANTS (AND CARBON-13 CHEMICAL SHIFTS) OF MISCELLANEOUS π SYSTEMS

J_{CC}(Hz) to labeled carbon (δ_{C-13}, ppm) of carbon no.:

Compound	1	2	3	4	5	6	7	8	9	10	11	12	Reference
118	1.37 (129.4)	2.56 (127.8)	0 (125.1)	0 (127.9)	0 (128.4)	0 (125.5)	0 (126.2)	0 (123.3)	3.17 (131.7)	0.77 (133.5)	57.04 (39.2)	(177.4)	2
119	45.1	2.83	4.30	0	0	0	0	2.93	2.15	2.54		54.93	2
120	47.36 (135.7)	2.93 (124.7)	3.42 (123.0)	0 (127.9)	0 (127.7)	0 (125.1)	0 (125.6)	1.34 (124.5)	2.69 (130.5)	2.69 (133.1)	(62.8)		2
121	13.7 (135.8)	8.4 (128.5)	0 (128.5)	0 (131.1)			42.7 (188.1)	(20.2)					2

122

2.68 (142.7)　1.70 (130.0)　4.70 (128.7)　1.0 (131.6)　2.68 (130.5)　1.10 (140.3)　0 (127.4)　0 (127.8)　0 (126.5)　　2

123

1.3 (129.1)　0.8 (127.0)　1.0 (128.0)　0.8 (128.7)　54.8 (53.5)　0.8 (36.2)　1.5 (137.3)　2.9 (135.5)　　7

124

2.4 (126.0)　0 (127.7)　0 (126.4)　0 (129.5)　(127.1)　1.3 (129.7)　4.3 (132.0)　1.9 (128.8)　　7

125

2.0 (135.0)　1.9 (129.9)　0.5 (128.9)　0.7 (127.3)　55.9 (41.1)　　30

167

TABLE 4-14. (continued)

| | J_{CC}(Hz) to labeled carbon (δ_{C-13}, ppm) of carbon no.: | | | | | | | | | | | | |
Compound	1	2	3	4	5	6	7	8	9	10	11	12	References
126	1.7 (140.8)	53.7 (137.3)	3.9 (135.6)	4.4 (139.1)	<1.0 (135.7)	3.9 (133.6)							30
149					13.95								8
151					7.9								8
152					8.6								8

168

Compound											Ref.
153 (NO₂ benzene)		7.6									8
160	(197.3)	40.2 (39.0)	1.1 (23.3)	2.7 (29.6)	3.3 (128.8)	(133.2)	3.6 (126.4)	1.2 (126.9)	50.9 (132.6)	2.9 (144.3)	26
161	(54.7)	(52.4)	1.0 (21.7)	2.8 (24.3)	5.9 (129.2)	(128.0)	5.3 (125.8)	2.9 (128.1)	56.0 (132.5)	(136.4)	26
162	(49.2)	37.9 (71.2)	1.0 (26.7)	1.9 (24.9)	2.9 (128.3)	(126.4)	3.6 (126.0)	2.0 (130.7)	43.4 (135.4)	1.5 (135.7)	26

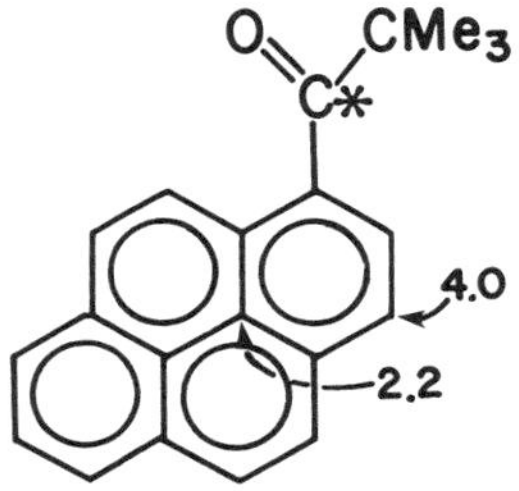

88

was observed above in the discussion of benzene compounds that twisting of the pivaloyl group attenuated couplings, and the same effect is seen in **88**. However, it is unclear why the two J values in **88** have now diverged, unless distortion about the 1-substituent places the labeled carbon closer to the peripheral coupling path.

Longer Range Couplings in Polynuclear Compounds

For one last time the reader should return to the list of carboxylic acids on p. 159. The only substantial longer range couplings are four-bonded couplings within the same ring, which are dependably ca. 1 Hz. A slightly larger value (1.3 Hz) is observed in 9-anthroic acid (**124**), where a dual path is present from the —CO_2H to the 10 position.

Longer range couplings outside the ring bearing the labeled substituent are usually not observed and in any event must be very small. The only measured couplings in the list of carboxylic acids are a ^4J value of 0.35 Hz and a ^5J value of 0.50 Hz in pyrene-1-carboxylic acid (**86**).[23]

For all functionalities and all aromatic systems, these trends continue: ^4J is about 1 Hz within the same ring, and all other couplings are small and may emerge as a measurable value in an unpredictable manner. The largest such coupling in this second category is a ^4J value of 0.54 Hz to the 5 position in methyl 1-naphthoate (**55**). The longest range coupling yet measured is to the 5

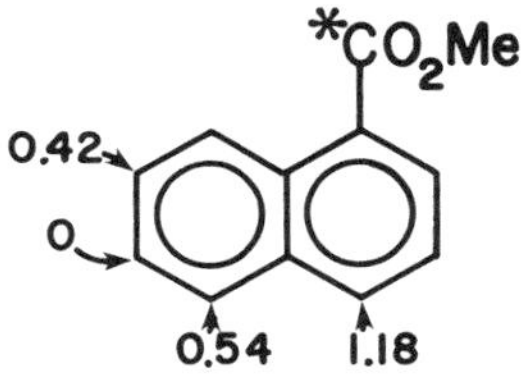

55

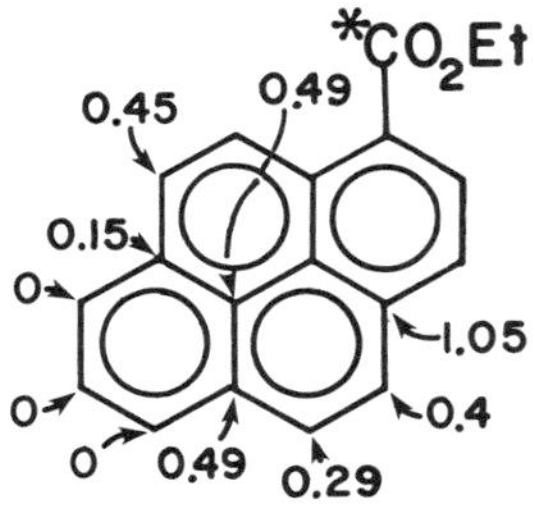

87

position in ethyl pyrene-1-carboxylate (**87**), for which a ^6J value of 0.29 Hz has been reported.[23]

Signs of Carbon–Carbon Couplings

A few sign-determination studies have been performed for nonconjugated systems and appear to reflect consistent trends. For α-substituted benzenes (see **137**), ^2J is positive, ^3J is positive, and ^4J is negative.[6] Most probably these signs can be assumed to hold with confidence for couplings in polynuclear systems.

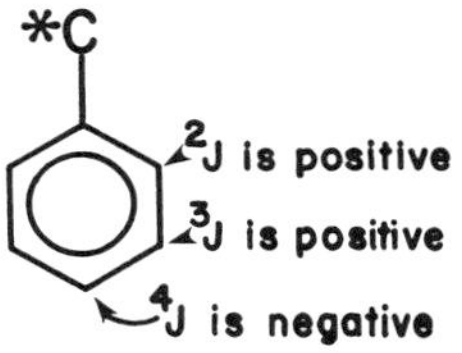

^2J is positive
^3J is positive
^4J is negative

137

Couplings in Conjugated Systems

Couplings in totally conjugated systems are larger than in nonconjugated ones and rely heavily upon the π-bond order of the associated bonds. The transmission of the coupling is more efficient in conjugated systems, and substantial couplings may occur with coupling pathways of four, five, or even six bonds.

An introduction to conjugated couplings is afforded by the hydrocarbons given below. Couplings to equivalent carbons are generally not given, because splittings between magnetically equivalent nuclei are not observed. Signs of J values are included (where available), because of the interesting sign behavior of the geminal couplings.

Signs of Carbon–Carbon Couplings

It was seen above that in α-labeled aromatics the signs of ^2J and ^3J were positive and of ^4J was negative. However, for conjugated systems, signs for geminal couplings may be positive or negative; for example, when geminal J values are compared, "larger" or "smaller" values may not be meaningful except in an algebraic sense. Inspection of **65**, **66**, **71**, **83**, and **84** demonstrates this variability of signs for the geminal couplings. Generally, geminal couplings to another ring are positive and within one ring are negative or near zero. To demonstrate coupling behavior in different rings, in anthracene (**71**) the geminal coupling between C-9 in the center ring and C-1 in the outer ring is $+1.80$ Hz (positive). To show behavior within the same ring, in phenanthrene (**83**) the geminal coupling between C-9 and the quaternary carbon in the same ring is -1.6 Hz (negative), and in naphthalene (**66**) the coupling between C-1 and C-10 is 0.20 Hz (small) and between C-1 and C-3 is -2.45 Hz (negative). In contrast to the variable behavior of geminal couplings, vicinal and four-bonded couplings behave consistently—the former are positive and the latter are negative, as was the case for nonconjugated systems.

Vicinal Couplings in Conjugated Systems

Vicinal couplings in the hydrocarbons presented above show a wide variability, from 3.05 Hz in pyrene (**84**) to 9.05 Hz in butadiene (**6**). A possible reason for this diverse behavior is the π-bond order of the associated bonds. Structures **138–143** give π-bond orders for the hydrocarbons[29] for which the J values are listed above. A correlation does emerge, but it is rough. In butadiene (**6**) and toluene (**65**) the ^3J values are 9.05 and 9.45 Hz, respectively, and the associated bond orders are likewise relatively large. In naphthalene (**66**) a smaller ^3J value to C-7 of 5.51 Hz is seen with lower associated bond orders. In anthracene (**71**) three different ^3J values exist—5.97 Hz to C-2, 3.13 Hz to C-4, and 7.6 Hz to C-10. The associated bond orders to C-2 are larger (0.61, 0.54, 0.74) than to C-4 (0.61, 0.48, 0.54), in agreement with these ^3J values, but the bond orders to C-10 are not large (0.61, 0.48, 0.61) and yet a larger ^3J value of 7.6 Hz is seen. One wonders at this point if the dual-path coupling to C-10 accounts for the larger coupling. In phenanthrene (83) four different ^3J values and in pyrene (84) three different ^3J values are seen, all of which appear to be dependent upon the associated π-bond orders. For example, the very small ^3J value in pyrene (3.05 Hz) has associated bond orders of 0.59, 0.52, and 0.54, whereas the larger ^3J value of 5.50 Hz in phenanthrene has associated bond orders of 0.77, 0.51, and 0.58. Cisoid couplings in these compounds tend to be quite large, again suggesting dual-path contributions. It is difficult to evaluate the relative contributions of each of the three bond orders in a vicinal linkage. Specifically, it is not obvious to what extent a high bond order can compensate for a low bond order in the same linkage. Attempts to correlate ^3J and bond orders have entailed the sums of the three bond orders of a vicinal linkage[19, 21] with questionable success.[14]

Instead of using the bond orders portrayed in 138–143, Hansen et al.[14] have used "generalized" bond orders as generated by a HMO computer program published for hydrocarbons.[29] It was found that these "three-bonded" bond orders (P_3) correlate well with both cisoid and transoid C—C couplings and also butadiene. Structures 144–148 demonstrate this correlation. For each structure the ^3J value is given with the corresponding "three-bond" bond order P_3. The correlation proceeds nicely through 144–148 if one takes the absolute value of P_3. For example, butadiene (144), with a large ^3J value of 9.05 Hz, has a corresponding large absolute P_3 value of 0.45. Pyrene (148) demonstrates a wide spread of both ^3J and P_3 (7.71 Hz, 0.28; 5.82 Hz, 0.21; 3.05 Hz, 0.03). In this last

5.51
$(P_3 = -.17)$

9.05
$(P_3 = -.45)$

144

145

5.97
$(P_3 = -.20)$

7.6
$(P_3 = -.40)$

3.13
$(P_3 = +.10)$

146

5.35
$(P_3 = -.13)$

5.85
$(P_3 = -.24)$

3.0
$(P_3 = +.04)$

147

5.82
$(P_3 = -.21)$

7.71
$(P_3 = -.28)$

3.05
$(P_3 = -.03)$

148

value of P_3 it is seen that P_3 is about zero, and thus a "residual" ^3J remains when P_3 vanishes. This led Hansen et al.[14] to propose the equation

$$^3J = 2.7 + 15.2 \times |\pi\text{-bond order}|$$

which he suggested separated the σ and π contributions to the vicinal coupling. It is remarkable that such a successful correlation is possible for the wide variety of olefinic, cisoid aromatic, and transoid aromatic couplings, particularly when the physical interpretation of this long-range generalized bond order is not clear. These bond orders are generated by the multiplication and summing of the appropriate coefficients of the linear combinations of the two atoms in question, and it is not obvious how the generated number describes the efficiency of π coupling throughout the route between the two atoms. It is also not clear why positive and negative values of this long-range π-bond order should not contribute in an algebraic manner rather than in an absolute sense. Nevertheless, Hansen's correlation has enjoyed a large measure of success, and the employment of these long-range bond orders might continue to be useful.

Substituent Effects

The effect of a hydroxy or alkoxy substituent is demonstrated by compounds **67, 73, 78, 85**, and **70**. Signs, when given, are those predicted by analogy with the parent hydrocarbons.[17] Because signs of these substituted compounds are not known with certainty, the trends to be discussed here have a degree of uncertainty, but they appear to be correct. First, geminal couplings appear to become less negative in the same ring and less positive in different rings. For example, the negative ^2J value of -2.45 Hz in naphthalene (66) becomes near zero (less negative) in 1-naphthol (67), and the positive ^2J value of $+1.80$ Hz in anthracene (71) becomes less positive ($+1.20$ Hz) in 9-methoxyanthracene (73). Accordingly, it was deduced that the negative ^2J value of -0.7 Hz in anthracene (71) to C-11 must become less negative and the corresponding ^2J value in 9-methoxyanthracene (73) of 1.48 Hz must be positive. Next, vicinal couplings are generally attenuated by the hydroxy substituent by a few tenths of a Hertz, but there is an exception in 1-hydroxypyrene (85; the coupling to the quaternary carbon is increased from 3.05 to 3.90 Hz). Most certainly all vicinal couplings are positive and no sign reversal occurs. Finally, four-bonded couplings are generally not affected much by the hydroxy substituent; the notable exception is the ^4J value of -1.47 Hz to C-5 in naphthalene (66) that changed to near zero in 1-naphthol (67).

The hydroxy and methoxy substituents in the compounds above modify the bond orders within the molecules, and extrapolations to other substituents would be difficult. Fortunately, a series of 1-labeled phenanthrene compounds 77–82 (Table 4-6) has been studied wherein the substituent has been varied considerably. The parent hydrocarbon itself was not studied, but 1-methyl-phenanthrene (80) should suffice for this purpose because the methyl group should have little effect on the couplings (a 9-methyl substituent in anthracene affects the couplings little, except for the geminal coupling to the 1-carbon, which would be susceptible to through-space effects: see 71 and 72 in Table 4-5).

This series amply demonstrates that in conjugated systems substituent effects are not as predictable as in nonconjugated (and particularly aliphatic) systems. Compounds **80, 77, 81,** and **82** exemplify the behavior of substituents. (The positions to which no J value is affixed have couplings of less than 0.4 Hz.) Perusal of these four examples permits no obvious correlations to be made. The electronegative substituents OH, CHO, and CN sometimes increase and sometimes decrease vicinal couplings, although the —CN group always increases ^3J. With regard to geminal couplings, the electronegative substituent

sometimes modifies ^2J and sometimes has little effect. The four-bonded couplings are little affected except when the substituent is —CN. Longer range couplings appear for the hydroxy compound **77** but not for the other compounds. This hodgepodge of couplings should be sufficient to serve as a caveat—in conjugated systems the effect of substituents is not yet understood. Gross trends, as portrayed in the parent hydrocarbons **6, 65, 66, 71, 83,** and **84** certainly exist; however, fine tuning is difficult.

References

1. Marshall, J. L.; Miiller, D. E. *Org. Magn. Reson.*, **6**, 395 (1974).
2. Marshall, J. L.; Faehl, L. G.; Kattner, R. *Org. Magn. Reson.*, **12**, 163 (1979).
3. Marshall, J. L.; Faehl, L. G.; Kattner, R.; Hansen, P. E. *Org. Magn. Reson.*, **12**, 169 (1979).
4. Becher, G.; Lüttke, W.; Schrumpf, G. *Angew. Chem. Intl. Ed. Engl.*, **12**, 339 (1973).
5. Maciel, G. E. In "NMR Spectroscopy of Nuclei Other Than Protons," Axenrod, T.; Webb, G. A. Eds.; Wiley-Interscience: New York, 1974, pp. 187–218.
6. Hansen, P. E. *Org. Magn. Reson.*, **11**, 215 (1978).
7. Marshall, J. L.; Faehl, L. G.; Ihrig, A. M.; Barfield, M. *J. Am. Chem. Soc.*, **98**, 3406 (1976).
8. Weigert, F. J.; Roberts, J. D. *J. Am. Chem. Soc.*, **94**, 6021 (1972).
9. Hansen, P. E.; Poulsen, O. K.; Berg, A. *Org. Magn. Reson.*, **7**, 405 (1975).
10. Linde, S. Aa.; Jakobsen, H. J. *J. Am. Chem. Soc.*, **98**, 1041 (1976).
11. Marshall, J. L.; Miiller, D. E.; Dorn, H. C.; Maciel, G. E. *J. Am. Chem. Soc.*, **97**, 460 (1975).
12. Ihrig, A. M.; Marshall, J. L. *J. Am. Chem. Soc.*, **94**, 1756 (1972).
13. Hansen, P. E.; Poulsen, O. K.; Berg, A. *Org. Magn. Reson.*, **9**, 649 (1977).
14. Hansen, P. E.; Poulsen, O. K.; Berg, A. *Org. Magn. Reson.*, **12**, 43 (1979).
15. Marshall, J. L.; Ihrig, A. M. *Org. Magn. Reson.*, **5**, 235 (1973).
16. Hansen, P. E.; Berg, A. *Org. Magn. Reson.*, **8**, 591 (1976).
17. Hansen, P. E.; Berg, P. *Org. Magn. Reson.*, **12**, 50 (1979).

18. Marshall, J. L.; Ihrig, A. M.; Miiller, D. E. *J. Mol. Spectrosc.*, **43**, 323 (1972).
19. Berger, S.; Zeller, K. P. *Chem. Commun.*, 423 (1975).
20. Hansen, P. E.; Poulsen, O. K.; Berg, A. *Org. Magn. Reson.*, **7**, 475 (1975).
21. Marshall, J. L.; Ihrig, A. M.; Miiller, D. E. *J. Magn. Reson.*, **16**, 439 (1974).
22. Berger, S.; Zeller, K. P. *Org. Magn. Reson.*, **11**, 303 (1978).
23. Hansen, P. E.; Poulsen, O. K.; Berg, A. *Org. Magn. Reson.*, **7**, 23 (1975).
24. Hansen, P. E.; Poulsen, O. K.; Berg, A. *Org. Magn. Reson.*, **8**, 632 (1976).
25. Buchanan, G. W.; Selwyn, J.; Dawson, B. A. *Can. J. Chem.*, **57**, 3028 (1979).
26. Cox, R. H. *Org. Magn. Reson.*, in press.
27. Evrard, G.; Piret, P.; Germain, G.; Van Meerssche, M. *Acta Cryst. B*, **27**, 661 (1971).
28. Grimaud, M.; Pfister-Guillouzo, G. *Org. Magn. Reson.*, **7**, 386 (1975).
29. Heilbronner-Straub, E. "Hückel Molecular Orbitals"; Springer Verlag: New York, 1966.
30. Marshall, J. L.; Faehl, L. G.; Ledford, N. D. *Spectrosc. Lett.*, **9**, 877 (1976).

5

SPECIAL TOPICS—SIGNS, ADDITIVITY OF J_{CC} VALUES, ALIPHATIC $^2J_{CC}$ AND $^4J_{CC}$ VALUES, AND CARBON/PROTON CORRELATIONS

Signs of Carbon–Carbon Coupling Constants

Coupling constants may be either positive or negative, depending on whether the more stable orientation of the two nuclear spins is antiparallel (positive) or parallel (negative).[1] For coupling patterns where the chemical shift differences are large relative to the coupling constants, the appearance of the spectrum is independent of the relative signs of the couplings. This straightforward spectral pattern is termed "first order," with splittings being essentially equivalent to the coupling constants. In contrast, if the chemical shift differences between nuclei are small relative to the coupling constants, the spectral pattern becomes "second order," with spacings not necessarily equal to coupling constants. In second-order spectra the intensities of lines change so that simple symmetrical patterns are commonly destroyed, and weaker "combination" lines appear at unusual positions in the spectrum. Signs of coupling constants can be quite important in determining a final second-order spectrum, with the placement and/or intensities of lines dependent upon the relative signs of couplings between nuclei that are tightly coupled.

Generalizations are difficult for second-order spectra. Obviously, the more tightly coupled nuclei are, the more bizarre the final pattern can be. A rule of thumb is that if $\Delta\delta/J < 10$, i.e., if the chemical shift difference between nuclei is

178

less than 10 times the coupling constant between them, second-order patterns can develop. General cases can be developed for simpler systems and have been described in many reference books.[2] For a more complicated spectrum with a larger number of diverse nuclei, the specific pattern must be uniquely analyzed by means of the routine LAOCOON program[3] which is now also available as a software package on newer generation NMR spectrometers.[4]

The importance of knowing the signs of coupling constants is apparent when analyzing second-order spectra, because the correct combination of relative signs is necessary for matching calculated and observed NMR spectra. Another reason for knowing signs, however, is the proper theoretical treatment of couplings. In this monograph the discussion of signs of couplings has been minimal, because the empirical determination of C—C couplings entails experimental difficulties not customarily encountered with more common nuclei, notably protons. Some discussion regarding signs of C—C coupling was warranted in Chapter 4, where conjugated geminal couplings could be negative or positive. In this chapter a more complete analysis of signs of J_{CC} values is made. Methods for determination of these signs are described, followed by a consideration of the known signs of J_{CC} for all types of couplings.

Determination of Signs of Couplings: General Aspects

Two general methods exist for the determination of signs of couplings. First, second-order patterns can sometimes give relative J signs if one particular combination of signs gives a satisfactory fit while other combinations do not. Figures 5-1 and 5-2 show examples where this method is applicable.[5] In these figures the ortho, meta, and para regions, respectively, of the proton NMR are shown for carboxyl-^{13}C-benzoate ester (1). The patterns are the AA'BB'C portion of a AA'BB'CX system where AA' = ortho protons, BB' = meta protons, C = para proton, and X = labeled carbon (see 2). A tightly coupled system exists here, and slightly different patterns are seen for the different respective sign

combinations of J_{AX}, J_{BX}, and J_{CX} (all J_{HH} values are positive). Inspection of these patterns will alert the reader to the possibility of subtle differences and to the dangers of drawing conclusions on the basis of best fits. A constant risk in this method of analysis is not knowing that indeed the best fit for each case has

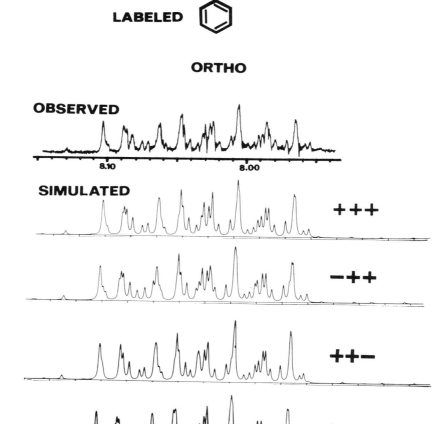

Figure 5-1. ^{1}H NMR of ^{13}C-*carboxyl*-methyl benzoate, ortho region.

been established by means of a sagacious and thorough search for all possibilities. It is possible, therefore, that the best "fit" may fortuitously result from the wrong combination of signs. The lesson to be learned is that a spectrum must be neither too simple nor too complex for this method of sign determination to succeed.

The second method of determining relative signs of coupling constants is by selective irradiation techniques. In this method specific transitions (lines) are perturbed with the concomitant observation of other transitions that will be

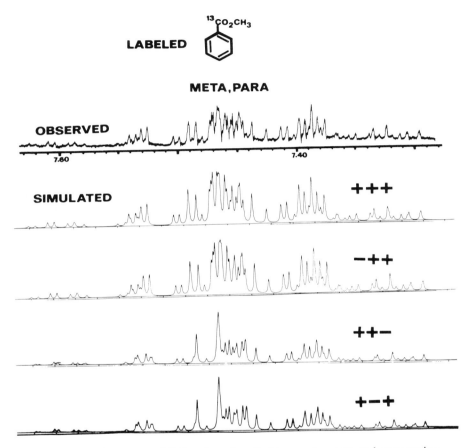

Figure 5-2. ^{1}H NMR of ^{13}C-*carboxyl*-methyl benzoate, meta and para region.

affected in a predictable manner depending upon the relative signs. Various irradiation methods are available, and in carbon couplings spin-tickling, spin-decoupling, and selective population transfer (SPT) techniques have been employed. A general discussion of these procedures exists[6] and only a cursory account is given here.

In spin tickling one resonance is weakly irradiated, leading to slightly more complicated patterns as other resonances are perturbed and are split into doublets. Analysis of the observed perturbations utilizes an energy level diagram[7] to ascertain the relationships between various nuclear spin transitions (i.e., resonances) for various sign possibilities. This analysis involves the following steps: (1) The normal NMR spectrum is studied by LAOCOON for all possibilities of sign combinations. For these different sign possibilities, the transition numbers generated by the LAOCOON program will be different. (2) The appropriate energy level diagram is set up to establish relationships between certain resonances. (3) Spin-tickling experiments are conducted to ascertain which resonances are in fact related. These relationships decide which sign combination is correct. A detailed explanation of the spin-tickling procedure has appeared.[8]

In spin-decoupling experiments several resonances are simultaneously strongly irradiated, giving rise to the collapse of multiplets in another portion of the spectrum to singlets. The energy level diagram[7] may be used here also to predict which multiplets collapse upon certain sign combinations. A classic example of this method in the literature[6] is the analysis of the 12-line pattern of an AMX system. In an AMX pattern each of the three signals from A, M, and X is split into a doublet of doublets. If the low (or high) doublet of A is irradiated and the low (or high) doublet of M coalesces to a singlet, then J_{AX} and J_{MX} are of the same sign, but if the high (or low) doublet of M coalesces, J_{AX} and J_{MX} are of different signs.

In certain experiments the level populations may be changed leading to observed spectra with intensities of various resonances changed and even inverted. This, the selective population transfer (SPT) experiment, can be used to determine relative signs by a judicious choice of resonances to be irradiated. The resulting spectral patterns are perturbed by certain rules which have been well formulated.[9] This method is best adapted to first-order spectra.

Symmetrical Double-Labeling Method (SDL)[10−14]

As pointed out above, in analyzing a second-order spectrum to determine signs, a compromise should be sought between patterns that are neither too simple nor too complex. An ideal choice, for example, would be an AA'B system, whose analysis is quite straightforward. The B portion of an AA'B system is a triplet and the AA' portion can be a strong pair of doublets with four weak absorption bonds,[2] although in practice generally the AA' portion appears as a doublet. Of concern to us in the SDL method is the analysis of the B triplet whose outer members encompass the algebraic sum $J_{AB} + J_{A'B}$. If J_{AB} and $J_{A'B}$ are of the same signs, then these outer members are separated by the distance $|J_{AB}| + |J_{A'B}|$. If J_{AB} and $J_{A'B}$ are of opposite signs, then the triplet contracts to reflect the quantity $|J_{AB}| - |J_{A'B}|$.

To observe the AA'B pattern, a symmetrically labeled compound must be synthesized with the labels occupying the A and A' positions, such as bibenzyl (3). The natural abundance B carbon signal is studied to measure the splitting of the triplet. For this method to work, the individual values J_{AB} and $J_{A'B}$ must be determined from study of monolabeled compound 4. In these examples 3 and 4 the quaternary carbon is identified as B, but each of the other carbons can be studied in turn. This method succeeds if B couples with both A and A'. Obviously systems should be chosen where A and A' strongly couple, for if

$J_{AA'} = 0$ the B portion of the spectrum will simply be overlapping doublets of J_{AB} and $J_{A'B}$.

Tables 4-5, 4-8, and 4-9 (Chapter 4) list couplings determined by the SDL technique. It is to be noted that each of the compounds listed in these tables is bilaterally symmetrical. Although only one carbon is shown to be labeled in each of these compounds, the actual compounds studied possessed two labeled carbons directly bonded about a plane of symmetry.

It must be remembered that this method involves the determination of the relative signs of J_{AB} and $J_{A'B}$ and success requires the knowledge of the absolute sign of one of these couplings. For C-3 in **5** (**89** in Table 4-8) this is easy, because the directly bonded coupling J_{AB} must be positive.[15, 16] Because J_{AB} and $J_{A'B}$ were determined to be of different signs, the ^2J value of 1.9 Hz must be negative. For C-4 in the same molecule (see **6**) more of an ambiguity exists, but one can

with confidence assign the vicinal coupling as positive.[17] Because J_{AB} and $J_{A'B}$ are of like signs, this ^2J value must be positive. This assumption that vicinal couplings are positive appears to be quite secure—no contradictions have ever arisen from this tenet in all of the J sign determinations.

Various trends emerge from Tables 4-8 and 4-9. These trends are exemplified by structures **7–20**. First, geminal aliphatic couplings are normally negative but are susceptible to electronic and steric effects. Accordingly, the ^2J value in **7** is -1.9 Hz, phenyl substituents afford a positive increment (**8**), and hydroxyl substituents more of a positive contribution (**9**). These positive contributions were noted to occur also for C—H and H—H couplings (see The Correlation of J_{CH} and J_{HH}: The J_{CH}/J_{HH} Ratio, Chapter 2). Structure **10** is an intermediate case where the role of the strained epoxide functionality is unclear. Negative geminal couplings also have been reported for the C—O—C moiety in CH_3OCOCl[18] (-2.8 Hz) and CH_3OCH_3[16] (-2.4 Hz).

Next, geminal couplings between a substituent carbon and an internal aromatic carbon are positive (**11–14**). Substituent effects on the terminal carbon produce relatively little effect on $^2J_{CC}$ and are in accord with similar effects in C—H couplings (see General Aspects of Aliphatic C—H Couplings, Chapter 2).

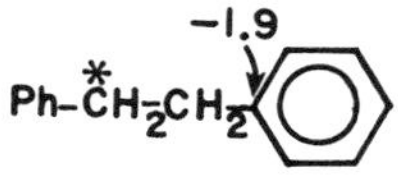

Ph–*CH₂CH₂–⟨ring⟩   −1.9

7

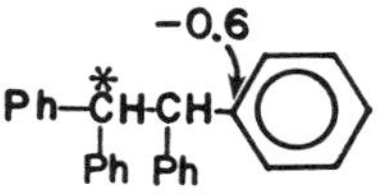

Ph–*CHCH–⟨ring⟩   −0.6
      | |
      Ph Ph

8

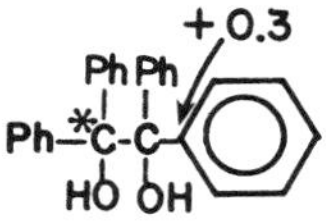

      Ph Ph   +0.3
      | |
Ph–*C–C–⟨ring⟩
      | |
      HO OH

9

      O
      ‖
Ph–*C–C–⟨ring⟩   −0.5
      |
      Ph Ph

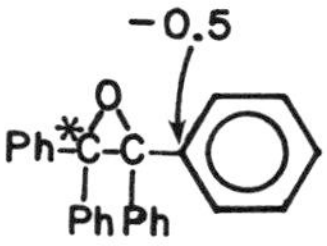

10

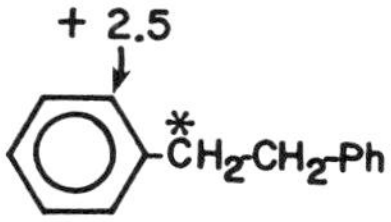

⟨ring⟩–*CH₂CH₂–Ph   + 2.5

11

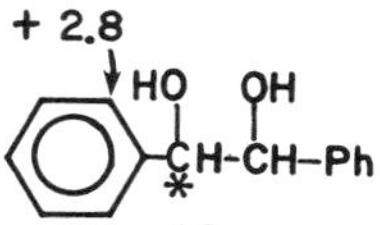

            HO   OH   + 2.8
            | |
⟨ring⟩–*CH–CH–Ph

12

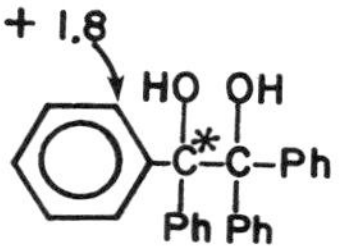

      HO OH   + 1.8
      | |
⟨ring⟩–*C–C–Ph
      | |
      Ph Ph

13

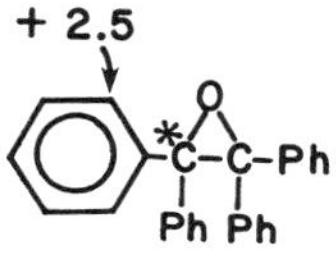

      O
      / \
⟨ring⟩–*C–C–Ph   + 2.5
      | |
      Ph Ph

14

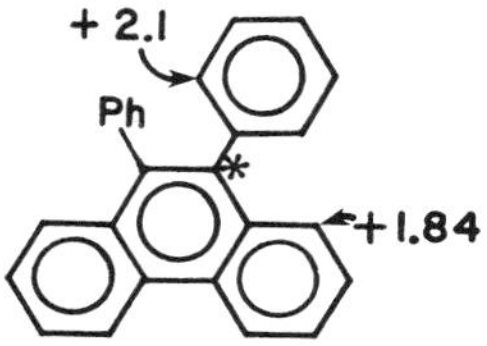

      Ph   + 2.1
      |
⟨phenanthrene system with *⟩   ⟵ +1.84

15

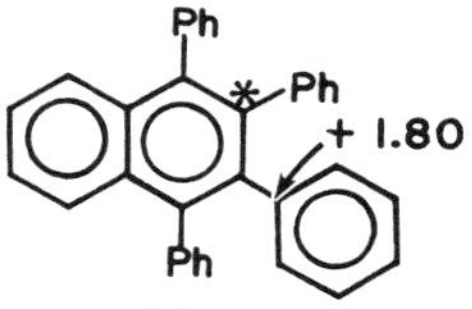

      Ph * Ph
      | /
⟨naphthalene system⟩   + 1.80
      |
      Ph   ⟨ring⟩

16

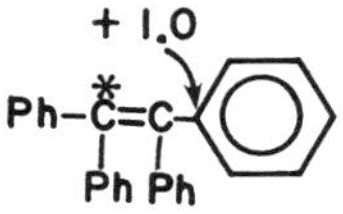

Ph–*C=C–⟨ring⟩   + 1.0
      | |
      Ph Ph

17

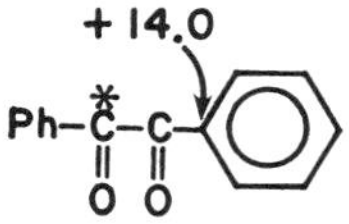

Ph–*C–C–⟨ring⟩   + 14.0
      ‖ ‖
      O O

18

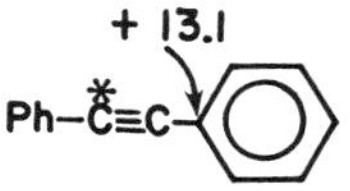

Ph–*C≡C–⟨ring⟩   + 13.1

19

(Br–⟨ring⟩)₂–*C=C–(⟨ring⟩–Br)₂   −0.7 ... +1.3

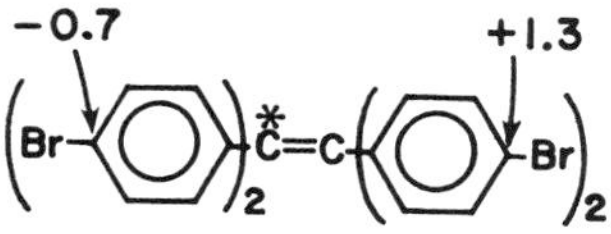

20

These couplings are positive also if the substituent is itself in another aromatic system (**15** and **16**)—it should be recalled from the discussion in Chapter 4 that ^{2}J values in polynuclear hydrocarbons are positive if the coupling nuclei are in different rings.

Geminal couplings through olefins (if measurable) are positive (**17**). Through a carbonyl, geminal couplings are very large and positive (**18**) and are in perfect analogy with geometrically equivalent H—H and C—H couplings (see **165** and **166**, in Chapter 2). Geminal couplings via an acetylene are likewise large and positive (**19**), as are geometrically equivalent C—H couplings (Table 2-11).

Little needs to be said regarding vicinal couplings determined by the SDL technique; they are all positive. Signs of four-bonded couplings are not as numerous as for ^2J or ^3J but appear consistently to be negative. In one example (20) a five-bonded coupling appears to be positive. This has been proposed as a generalization[13] for carbon–carbon couplings: ^3J is positive, ^4J is negative, and ^5J is positive, and the proposal has a theoretical basis.[17]

Several other compounds have been studied by the SDL technique, but discussion of these is deferred until the next section, because they involve dual coupling paths. These compounds are the phthaloyl compounds 50–53 (Table 4-2) and the anthracene compounds 74 and 76 (Table 4-5).

Irradiation Methods

Methyl tetrolate (21) has been analyzed by the spin-decoupling[19] and SPT[20] techniques. In the spin-decoupling experiment the trilabeled derivative (22) was synthesized and the spectrum was analyzed in a straightforward fashion by selective decoupling of the appropriate doublet of the AMX pattern in the ^{13}C NMR spectrum. Accordingly, all signs were determined as positive from the assumption that the directly bonded carbon was positive. Unfortunately, the acquisition of the data is not routine: particular hardware equipment is required to perform the specialized triple resonance experiment of acquiring ^{13}C data while simultaneously irradiating ^{13}C signals. Furthermore, the synthesis of the trilabeled compound was arduous. Another simpler approach was to synthesize the monolabeled compound 23 with subsequent analysis by the SPT method. Selective irradiation in the proton region was applied by pulses to various transitions of either methyl group. The results were in agreement with those of the spin-decoupling experiment. Additionally, signs were also determined for couplings not available by the spin-decoupling experiment for 22. The signs determined are listed in 24. The only assumptions in this analysis were that all

$$CH_3C\equiv C\text{-}CO_2Me \qquad \overset{*}{C}H_3\text{-}\overset{*}{C}\equiv C\text{-}\overset{*}{C}O_2Me \qquad CH_3\text{-}C\equiv C\text{-}\overset{*}{C}O_2Me$$

$$\underline{21} \qquad\qquad \underline{22} \qquad\qquad \underline{23}$$

directly bonded couplings were positive.

$$\underline{24}$$

Summary

A comparison of the three methods of sign determination for C—C couplings reveals advantages and drawbacks of each technique.

In the symmetrical double-label (SDL) method, the spectroscopic procedure is simple because no irradiation procedure is necessary. Instead, normal acquisition of the spectrum may be accomplished and analysis involves simply the measurement of splittings and comparison of these spacings with results obtained from monolabeled compounds. The drawback of this method is that suitably symmetrical systems must come from the drawing board; accordingly, the systems studied tend to be specialized. However, once a system is entered synthetically, derivatization allows a wide variety of functionalities to be made—in Table 4-8, for example, a wide gamut of sp^3, sp^2, and sp compounds was made available synthetically from the precursors.

The spin-decoupling method embodies a blockbuster approach. No consideration need be made of symmetry considerations or of complex proton couplings, as long as one chooses to tackle the tedious synthetic task of incorporating several labels in a molecule. The spectroscopic method requires specialized equipment, available only in particular NMR laboratories.

The SPT technique requires only a single ^{13}C label and the spectroscopic acquisition of data usually can be accomplished via routinely available software. The drawback is that specific irradiation of proton signals requires a clean, nonintricate pattern in the proton coupled ^{13}C NMR spectrum; hence, careful design of the molecule is necessary.

The Additivity of Carbon–Carbon Coupling Constants

The additivity of NMR coupling constants over more than one path has been implied in the literature,[21-27] and several examples exist of two simultaneous coupling pathways that appear to be anomalously large.[23, 25, 26, 28] For example, the dual homoallylic proton–proton coupling in 1,4-cyclohexadienes (25) is

usually large for a five-bonded proton–proton coupling (ca. 8–10 Hz)[29] compared with ca. 0–2 Hz for monopath compounds (26).[30] However, other examples exist where a dual-path coupling appears reasonably low.[31–33] An example is anthraquinone[34] (27, taken from Table 4-5) where the dual-cis coupling is 3.84 Hz compared with 1.4 Hz for dimethylphthalate (28, taken from Table 4-2); because of the orientation effect of the carbonyl (see $^3J_{CC}$ vs. Orientation of a Terminal Substituent, Chapter 3), the coupling in this latter compound 28 would be expected to be small, as observed. Further, in 9-anthrone[35] (29, taken from Table 4-5) the coupling is 3.28 Hz, not that much larger than the value estimated for the transoid conformation shown for

o-tolualdehyde (34) (see General Couplings in Polynuclear Compounds, Chapter 4). However, all of these systems had two identical pathways and symmetry complications are a possibility (i.e., it is clear that a dual-path coupling does not necessarily "double" the coupling[28, 36]).

In an unsymmetrical system, however, these complications cannot arise, and it should be possible to test the idea that the couplings are additive. Furthermore, if two pathways for an overall coupling each provide a coupling of different signs, then to determine whether or not the couplings are algebraically additive should be straightforward. Carbon–carbon J values are ideal for exploring this idea, because unlike the proton nucleus, carbon is capable of providing completely separate multiple coupling pathways (see 31).

Because geminal aliphatic couplings are negative and vicinal aliphatic couplings are positive, a likely candidate would be 32 with simultaneous two-bonded and three-bonded couplings. Accordingly, compounds 33 and 34 were synthesized and the J values were determined (taken from Table 3-2 and Table 4-1, respectively).[37] In 33 an observed ^{2+3}J of 1.22 Hz is much smaller than the anticipated cis-^3J value of ca. 4 Hz (see $^3J_{CC}$ vs. Orientation of a Terminal Substituent, Chapter 3) and is more like the anticipated geminal value of 0–2 Hz. Hence, an algebraic addition of a positive ^3J and negative ^2J is

33   34   35   36

occurring, giving rise to an overall reduced splitting. In **34** two different ^{2+3}J values are observed, both of which are consistent with a algebraic addition of two couplings. The value of 1.71 Hz to C-4 in **34** is proportional to the value of 1.22 Hz in **33** and is again small for the same reason. By contrast, the J value to C-3 in **34** is larger because the geminal ^2J value through the olefin should be near zero (see Olefins and Acetylenes, Chapter 4). The observed value of 4.88 Hz is, therefore, essentially the vicinal coupling. Other five-membered rings with apparent algebraic addition are **35** and **36** (taken from Table 3-2) where the indicated couplings are quite small.[38] In **35** and **36** a different type of vicinal pathway is present with an internal ketone, but unfortunately a good model is

37

not available to suggest an accurate ^3J value for the *cis*-2-butanone moiety. Compound **37** (taken from Table 4-14) suggests a rather large value, but the ^3J value of a 2-butanone group frozen into a cis conformation is probably much less.[37] (More discussion on the ^3J contribution of a *cis*-2-butanone occurs later in this section).

There are examples of aromatic compounds that also possess simultaneous positive and negative coupling contributions, leading to unusual results. For example, in dihydroanthracene[14] (**38**, taken from Table 4-5) the two geminal couplings are quite different. The ^2J value to C-1 appears normal, for such

38

couplings should be positive and about 2–3 Hz (see Table 4-2). By contrast, the J value to C-11 is near zero, an unexpected result until it is realized that the four-bonded contribution (via the other aromatic ring) should be negative (see

Symmetrical Double-Labeling Method, this chapter), although it is surprising that this negative contribution is so large. In the carbonyl compounds **39** and **40** (taken from Table 4-5) the coupling to C-11 is larger, and presumably more

positive in both. It is to be recalled from compounds **109** and **111** (Chapter 4, Geminal Couplings in Polynuclear Compounds) that an orientation effect of the carbonyl should magnify the couplings to C-11 by as much as 2 Hz. That the actual difference between the two J values in **39** and **40** is only ca. 1 Hz may reflect again a negative four-bonded contribution to C-11.

The concept of additivity of J values in five-membered rings can be used to elucidate geminal signs in systems containing heteroatoms. Because geminal sp^3—sp^3 couplings through oxygen are negative (e.g., **145**, Table 3-2) one might anticipate the indicated coupling in phthalic anhydride (**41**, taken from Table 4-2) to be attenuated. This coupling is actually large and presumably positive (5.6 Hz) and in phthalimide (**42**) is even larger (8.8 Hz). This suggests that the

geminal couplings through heteroatoms between carbonyls (i.e., OC—X—CO) are positive, and model couplings involving protons support this view.[14] Furthermore, in the lactone **43** and lactam **44** (taken from Table 4-11) couplings of 3.57 and 6.57 Hz, respectively, are seen to C-8, even larger than **45** which has two vicinal contributions to C-8. The data therefore indicate carbonyl-heteroatom–carbonyl geminal couplings are positive (and are larger through nitrogen than through oxygen). Once this is recognized, then aliphatic couplings in norbornane lactones and ethers can be rationalized. In compounds **46** and **47** (taken from Table 3-2) the larger value of 3.1 Hz for **46** can be viewed as the sum

of positive ^3J and ^2J values. In **47**, the geminal coupling is electronically like that in ethers and should be negative, and the smaller value of ca. 0.5 Hz is the algebraic sum of a positive vicinal coupling and a negative geminal coupling.

In other types of five-membered rings, two different contributions exist both of which are expected to be positive. In fluorene (**48**) and fluorenone (**49**, taken from Table 4-10) the indicated J values are sizable, and the particularly large

value of 7.88 Hz in **49** reflects the "double-orientation" effect of the carbonyl, which should enhance both the geminal and the vicinal positive contributions. In **50** (taken from Table 4-10) and **51** (taken from Table 4-11) the effect is even greater, giving couplings of 9.8 and 10.99 Hz, respectively, presumably because

aliphatic C—C—C—CO couplings can be larger than aromatic C—C—C—CO couplings. Also in **50** and **51** are the smaller dual path couplings of 3.66 and 2.65 Hz, respectively. For these smaller couplings the geminal contributions are aliphatic and should be quite small[38] (ca. 1 Hz) and may even be negative.

In compounds **52** (taken from Table 4-10) and **53**, **54**, and **55** (taken from Table 3-2) the geminal contributions across the carbonyl should be very large

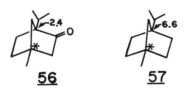

and positive (in acetone, Table 3-2, the geminal coupling across the carbonyl is +16.1 Hz, and in acetophenone it is 13.7 Hz, Table 4-14). These couplings are indeed large (15.5–17.58 Hz) and may reflect the sum of this large geminal coupling and the modest vicinal coupling of the C—C—CO—C moiety. Model compounds for this vicinal moiety are not readily available, but this vicinal value is probably quite small when the conformation is cis.[37] The other dual-path couplings in these compounds (0.6–1.56 Hz) are very small and even suggest that these J values may essentially reflect the geminal values with the vicinal couplings being near zero. A ^3J value of 2.4 Hz is known for **56** (taken from Table 3-2) in which a *cis*-butanone moiety exists, but of course there are also vicinal and geminal contributions. Comparison of this value of 2.4 Hz with that of 6.6 Hz in **57** (taken from Table 3-2) where the carbonyl has been

removed, however, again suggests that the vicinal coupling through the carbonyl is near zero.

In all of the discussion above, systems were analyzed that had different coupling contributions. In fact, care was taken to choose molecules that had simultaneous vicinal and geminal couplings, necessarily of quite different nature. However, several examples exist where simultaneous vicinal couplings exist, sometimes identical, and it is unclear how much the observed J values are augmented by equivalent pathways. For example, it has been noted at the beginning of this section that uncertainty exists whether the 1,4 couplings in aromatic rings should be considered "doubled" values of a single cis coupling. The difficulty is that good model compounds are not available—for example, what would be the single-path model compound for the 1,4 coupling in benzene?

For aliphatic compounds better single-path model compounds are available. In compounds **58** and **59**, and **60** and **61** (taken from Table 3-2), ^3J increases as an additional vicinal coupling route is affixed, but the values are not doubled (compare 5.42 and 8.49 Hz; 4.15 and 7.81 Hz). However, changing **58** to **59** and

58   **59**   **60**   **6l**

60 to **61**, should amplify a negative contribution to ^3J owing to a more compact molecule and to more C-1–C-3 orbital interactions (see Other Contributions to Carbon–Carbon Vicinal Couplings, Chapter 3). It is difficult to assess independently these various factors.

In compounds **62** and **63** (taken from Table 3-2) examples are seen with three vicinal pathways; in **63** all three are identical. These two observed J values of

62   **63**

12.7 and 12.2 Hz are about the same, and so substituting the olefin in **62** for the ethano bridge in **63** has little effect. When these J values are compared with single-path aliphatic couplings, one sees that they are clearly augmented. *cis*-Aliphatic couplings for the butane moiety would be perhaps 4–5 Hz (see Table 3-8), and at first sight it therefore appears that the observed couplings in **62** and **63** are about triple the value of a monopath coupling. However, no terminal hydrogens exist on the cis moiety in **62** and **63** and through-space interactions normally involving these hydrogens (see Other Contributions to Carbon–Carbon Vicinal Coupling, Chapter 3) are absent. These through-space contributions are positive, and the result would be a much reduced vicinal J value: for the *cis*-butane moiety, it has been calculated[39] that removal of these interactions should reduce the vicinal J value to as low as 2 Hz. The question thus emerges, why are the couplings in **62** and **63** not smaller?

In the norbornane compound **57** a value of 6.6 Hz was noted between the bridgehead carbons. This J value is intermediate between the "triple" J in **62**, **63**, and "normal" monopath cis couplings and therefore a qualitative approach to evaluating multipath couplings appears to work in aliphatic couplings. In **57** a geminal contribution across the methylene bridge C-7 also exists, which has been seen before to be substantial in some cases and should be negative. The coupling in **57** therefore may be a composite between two positive vicinal couplings and a negative geminal contribution. Without the geminal contribution, the J value in **57** might have been larger. It is tempting to suggest that without this geminal coupling the observed J in **57** would be more nearly two-thirds the J value in **62** and **63**.

In summary, it is apparent that carbon–carbon couplings, at least roughly, are algebraically additive when more than one path is present. The attenuation of observed couplings when positive vicinal and negative geminal couplings are simultaneously present particularly demonstrate this phenomenon. In some cases, it appears that relative values, and even signs, of couplings can be deduced by "subtracting" anticipated contributions from the observed J values. The problem with any of these arithmetical manipulations is that model compounds may not be available to furnish accurate J values for some of the contributions in question. Prudence is therefore dictated in any sort of analysis involving multipath couplings.

Two-Bonded and Four-Bonded Aliphatic Couplings

Chapter 3 was mostly concerned with vicinal carbon–carbon couplings in aliphatic systems, although cursory inspection was made of geminal (^2J) and longer range (^4J) couplings. Not a great deal of attention has been devoted to aliphatic ^2J and ^4J values in the literature, because these values are commonly quite small and frequently below the limits of resolution of the spectrometer. This situation is to be contrasted with that concerning aromatic ^2J and ^4J values, which were seen in Chapter 4 to be in general quite substantial. Further inspection is warranted, however, of these ^2J and ^4J aliphatic values in an effort to see whether recognizable trends exist. For example, a closer look should be made of geminal couplings to see whether a dependence exists of ^2J upon the orientation of an electronegative substituent, as was noted for carbon–proton couplings (see Geminal J Values in Conformational Analysis, Chapter 2). As an another example, it would be of interest to observe whether four-bonded carbon–carbon couplings display the "W" coupling, well known in norbornanes and bicyclo[2.1.1]hexanes as a dependable, stereospecific coupling in proton–proton systems that possess the proper "zigzag" geometry.[30]

Geminal Couplings

As noted briefly in Chapter 3, geminal carbon–carbon couplings in aliphatic compounds are typically small and are commonly measured as "zero," i.e., below the resolution limits of the spectrometer. As one peruses the geminal coupling of Table 3-2, one notices that these geminal couplings may vary considerably, from ca. 0 to 3 Hz, in an apparently random fashion. A clue to the reason for this behavior may be afforded by **64** and **65** (taken from Table 3-2). In

64   **65**

each of these compounds, two different geminal J values exist and the only difference between the two geometries in each compound is the orientation of the substituent. In each compound the smaller coupling is associated with the geminal carbon trans to the substituent. As further examples, **66** and **67** (taken from Table 3-2) each have different geminal couplings. The remarkable feature of the geminal couplings in **66** and **67** is the reversed relative values of the ^2J

values, i.e., in **66** the C-2 coupling is 1.68 Hz and the C-3 coupling is near zero, whereas in **67** the relative J values to C-2 and C-3 are exchanged. The reason for this behavior in ^2J values is clearly the different geometry in the secondary alcohol functionality. When the preferred geometry of the hydroxyl is anti to the coupled carbon (C-3 in **66**, C-2 in **67**), the coupling is small. The analogy of this behavior with that in the adamantane compounds **64** and **65** is striking.

The examples **64–67** strongly suggest a dependence of ^2J upon the orientation of a terminal substituent. It is to recalled from Chapter 2 (Geminal J Values in Conformational Analysis) that with regard to C—H couplings, a terminal electronegative substituent trans to the coupling path affords a positive contribution to $^2J_{CH}$, whereas the presence of such a substituent syn to the coupling path contributes a negative increment. This same effect has been suggested for C—C couplings.[40] However, as Hansen has observed,[40] "How far this simplified approach can be taken has still to be shown...". Until signs have been determined for geminal compounds with a substituent of fixed geometry, the uncertainty regarding this matter will not be removed. Until then, it must suffice to recognize that the variable behavior of geminal couplings of such compounds as **68–74** (taken from Table 3-2) may well reflect the effect of the orientation of electronegative substituents, including a consideration of the signs of the J values.

So far, the discussion involving geminal aliphatic C—C couplings has involved systems with a terminal electronegative substituent. A few examples exist for ^2J values in systems containing internal substituents; the paucity of data in this regard precludes the postulation of any trends or generalizations. Compounds **75** and **76** suggest that, if anything, such systems involve ^2J values that are substantial. The suggestion has been made that the couplings in these systems are more positive.[40] If no electronegative substituent is present, as in **77** and **78**, it is possible for the coupling either to be substantial or near zero (all ^2J values taken from Table 3-2).

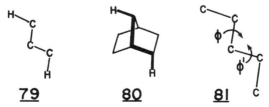

In summary, geminal aliphatic carbon–carbon couplings behave in a manner that superficially appears to be random, but contributions to ^2J exist that involve the orientation of a terminal substituent. These contributions may be either negative or positive in sign, although it must be recognized that sign determinations for $^2J_{CC}$ are few in number. Those which have been measured generally are negative. The potential use of $^2J_{CC}$ values for conformational analysis may be real, but useful applications must wait for further research developments.

Four-Bonded Couplings

In proton NMR, four-bonded couplings are infrequently resolvable and are of limited utility. One notable exception is the "W" coupling,[30] which is of some use in stereochemical elucidation of bicyclic and related systems (see **79**). In **80** this "W" geometry is portrayed within a norbornane system.

Enough examples exist of four-bonded couplings (from Table 3-2) to allow an examination of the dependence of $^4J_{CC}$ upon geometry. To describe the geometry of a four-bonded coupling, two dihedral angles ϕ and ϕ' can be used. For the "W" coupling in 79, $\phi = \phi' = 180°$. In compounds 82–92, a variety of four-bonded coupling geometries occurs. For example, in 83 $\phi = \phi' \approx 0°$ for the coupling between the —CO_2H group and the syn methyl at the C-7 methylene bridge, and in 87 $\phi = \phi' = 180°$ for the —CO_2H group and either of the methyl groups. The striking feature of the couplings in 82–92 is the lack of trends. In particular, the W geometry is not associated with the larger J values (in 84, $J = 0.31$ Hz; in 86, $J = 0.33$ Hz; in 87, $J = 0.29$ Hz; in 88, $J < 0.18$ Hz). The largest J in 82–92 is 0.73 Hz in 85 for a geometry of $\phi \simeq 0°$, $\phi' \simeq 0°$, suggesting a

through-space interaction between the coupling carbons that augments the coupling. In the carboxylic acid analog (83), however, the same coupling is small, possibly because of negative through-space contributions through the carbonyl group. Unfortunately, it appears that four-bonded carbon–carbon couplings cannot be used with confidence in relation to the associated geometries.

The Correlation of Carbon and Proton Coupling Constants

Implicit in our general discussion of carbon–carbon couplings was the idea that in systems where the carbon was contained in a substituent, carbon–carbon couplings could be related to geometrically equivalent carbon–proton or proton–proton couplings. For example, in the series **93–95** the geometrically equivalent geminal H—H, C—H, and C—C couplings are observed across a

carbonyl functionality.[23] The ratio J_{CH}/J_{HH} is 0.65 and the ratio J_{CC}/J_{CH} is quite similar, 0.62. In this series of compounds, therefore, the substitution of a proton by a carbon reduces the J value by a factor of 0.6–0.7. It is to be recalled from Chapter 2 (Tables 2-11 and 2-12) that the factor of ca. 0.4–0.8 was quite constant in relating H—H and C—H couplings.

In compounds **96–98** vicinal acetylenic couplings are compared (values taken from Tables 2-1 and 4-1). The ratios J_{CH}/J_{HH} and J_{CC}/J_{CH} are, respectively, 0.47 and 0.41. Compounds **99** and **100** compare geminal acetylenic compounds; here the ratio J_{CC}/J_{CH} is 0.40.

Compounds **101–103** compare vicinal olefinic couplings, from which are obtained respective J_{CH}/J_{HH} and J_{CC}/J_{CH} ratios of 0.66 and 0.20 for the cis

couplings and 0.74 and 0.53 for the trans couplings. Here the only ratio that appears out of line is the 0.20 J_{CC}/J_{CH} value for the cis couplings, but it is to recalled from Chapter 3 ($^3J_{CC}$ vs. Orientation of a Terminal Substituent) that through-space interaction via the carbonyl reduces the cis J_{CC} value.

Compounds **104–106** represent vicinal aliphatic couplings. No doubt torsional effects will change the geometry of the coupling nuclei in **106** relative to **104** and **105**, but the comparison is interesting. The respective J_{CH}/J_{HH} and J_{CC}/J_{CH} ratios are 0.50 and 0.46 for the cis couplings and 0.68 and 0.39 for the trans couplings.

For compounds **103** and **106** no sign was indicated for the vicinal couplings but most certainly they are positive. Therefore, all ratios of J_{CC}/J_{CH} and J_{CH}/J_{HH} are positive and lie in the total range of 0.20–0.74. The J_{CC}/J_{CH} ratio tends to be lower than the J_{CH}/J_{HH} ratio, and in the cis geometry (**103** and **106**) is

+9.10 (cis)   +4.54 (cis)   2.09 (cis)
+3.70 (trans)   +2.51 (trans)   0.98 (trans)

104   105   106

anomalously low owing to through-space interactions. However, other than this unique case, J_{CC}/J_{CH} generally falls in the range +0.40–0.80. It appears safe to conclude that carbon couplings operate via mechanisms similar to those of proton couplings. It has been noted in Chapter 3 (Other Contributions to Carbon–Carbon Vicinal Couplings) that additional mechanistic contributions may exist in carbon couplings, because quadrivalent carbon can bear more substituents. However, the similarities to proton couplings are sufficient to allow proton–proton or carbon–proton couplings to be used as models for carbon–carbon couplings.

References

1. Emsley, J. W.; Feeney, J.; Sutcliffe, L. H. "High Resolution Nuclear Magnetic Resonance Spectroscopy"; Pergamon Press: New York, 1965; Vol. 1, p. 162.
2. For example, see ref. 1, pp. 310–428.
3. Bothner-By, A. A.; Castellano, S. M. In "Computer Programs for Chemistry," Detar, D. F. Ed.; Benjamin: New York, 1968; Vol. 1, p. 10.
4. For a listing of this program, see Cooper, J. W. "Spectroscopic Techniques for Organic Chemists"; John Wiley & Sons: New York, 1980; pp. 339–355.
5. Ihrig, A. M.; Marshall, J. L. *J. Am. Chem. Soc.*, **94**, 3268 (1972). An example of this method using C—C couplings involved a glucosuranose derivative, where vicinal coupling were determined to be positive. Gagnaire, D.; Reutenauer, H.; Taravel, F. *Org. Magn. Reson.*, **12**, 679 (1979).
6. Hoffman, R. A.; Forsén, S. In "Progress in Nuclear Magnetic Resonance Spectroscopy," Emsley, J. W.; Feeney, J.; Sutcliffe, L. H. Eds.; Pergamon Press: New York, 1966; Vol. 1, p. 15.
7. Castellano, S. M.; Bothner-By, A. A. *J. Chem. Phys.*, **47**, 5443 (1967).

8. Marshall, J. L.; Seiwell, R. *Org. Magn. Reson.*, **8**, 419 (1976).
9. Sørensen, S.; Hansen, R. S.; Jakobsen, H. J. *J. Magn. Reson.*, **14**, 243 (1974); Jakobsen, H. J.; Linde, S. Aa.; Sørensen, S. *J. Magn. Reson.*, **15**, 385 (1974); Chalmers, A. A.; Pachler, K. G. R.; Wessels, P. L. *Org. Magn. Reson.*, **6**, 445 (1974).
10. Hansen, P. E.; Poulsen, O. K.; Berg, A. *Org. Magn. Reson.*, **7**, 405 (1975).
11. Hansen, P. E.; Poulsen, O. K.; Berg, A. *Org. Magn. Reson.*, **8**, 632 (1976).
12. Hansen, P. E.; Berg, A. *Org. Magn. Reson.*, **8**, 591 (1976).
13. Hansen, P. E.; Poulsen, O. K.; Berg, A. *Org. Magn. Reson.*, **12**, 43 (1979).
14. Hansen, P. E.; Berg, A. *Org. Magn. Reson.*, **12**, 50 (1979).
15. Grant, D. M. *J. Am. Chem. Soc.*, **89**, 2228 (1967).
16. Dreeskamp, H.; Hildenbrand, K.; Pfisterer, G. *Mol. Phys.*, **17**, 429 (1969).
17. Sardella, D. J. *J. Am. Chem. Soc.*, **95**, 3809 (1973).
18. Ziessow, D. *J. Chem. Phys.*, **55**, 984 (1971).
19. Marshall, J. L.; Miiller, D. E.; Dorn, H. C.; Maciel, G. E. *J. Am. Chem. Soc.*, **97**, 460 (1975).
20. Linde, S. Aa.; Jakobsen, H. J. *J. Am. Chem. Soc.*, **98**, 1041 (1976).
21. Barfield, M.; Spear, R. J.; Sternhell, S. *J. Am. Chem. Soc.*, **93**, 5322 (1971); Barfield, M.; Spear, R. J.; Sternhell, S. *J. Am. Chem. Soc.*, **97**, 5160 (1975); Barfield, M.; Chakrabarti, B. *J. Am. Chem. Soc.*, **91**, 4346 (1969).
22. Günther, H. *Tetrahedron Lett.*, 2967 (1967).
23. Weigert, F. J.; Roberts, J. D. *J. Am. Chem. Soc.*, **94**, 6021 (1972).
24. Lozac'h, R.; Braillon, B. *J. Magn. Reson.*, **12**, 244 (1973).
25. Garbisch, E. W. Jr.; Griffith, M. G. *J. Am. Chem. Soc.*, **90**, 3590 (1968).
26. Durham, L. J.; Studebaker, J.; Perkins, M. J. *Chem. Commun.*, 456 (1965).
27. Murrell, J. N. In "Progress in Nuclear Magnetic Resonance Spectroscopy," Emsley, J. W.; Feeney, J.; Sutcliffe, L. H. Eds.; Pergamon Press: New York, 1970; Vol. 6, p. 1.
28. Marshall, J. L.; Faehl, L. G.; McDaniel, C. R. Jr.; Ledford, N. D. *J. Am. Chem. Soc.*, **99**, 321 (1977).
29. Rabideau, P. W. *Acc. Chem. Res.*, **11**, 141 (1978).
30. Sternhell, S. *Quart. Rev.*, **23**, 236 (1969).
31. Berger, S.; Zeller, K.P. *Chem. Commun.*, 423 (1975).
32. Ihrig, A. M.; Marshall, J. L. *J. Am. Chem. Soc.*, **94**, 1756 (1972).
33. Marshall, J. L.; Ihrig, A. M. *Org. Magn. Reson.*, **5**, 235 (1973).
34. Hansen, P. E.; Poulsen, O. K.; Berg, A. *Org. Magn. Reson.*, **9**, 649 (1977).
35. Marshall, J. L.; Ihrig, A. M.; Miiller, D. E. *J. Magn. Reson.*, **16**, 439 (1974).
36. Rabideau, P. W.; Paschal, J. W.; Marshall, J. L. *J. Chem. Soc. Perkin Trans. 2*, 842 (1977).
37. Marshall, J. L.; Faehl, L. G.; Kattner, R. *Org. Magn. Reson.*, **12**, 163 (1979).
38. Berger, S. *Org. Magn. Reson.*, **14**, 65 (1980).
39. Barfield, M. *J. Am. Chem. Soc.*, **102**, 1 (1980).
40. Hansen, P. E. *Org. Magn. Reson.*, **11**, 215 (1978).

6

BIOCHEMICAL APPLICATIONS OF CARBON–CARBON COUPLINGS

Biosynthetic Applications

Biosynthesis

Study of biosynthetic pathways was aided greatly by the discovery of isotopes.[1] These isotopes could be used as tracers, or labels, to monitor the fate of an individual atom as it moved through a sequence of in vivo reactions. Radioactive nuclei, such as ^{14}C, ^{3}H, or ^{35}S, could be implanted into specific sites in molecules by tailor-made syntheses[2] and could be detected after the molecules had passed through a certain metabolic sequence in a plant or animal. Identification of the exact site of the label in the product was required in order to ascertain how the atoms had moved about. Unfortunately, this identification suffered from two drawbacks: (1) special safety precautions were mandated by the potentially dangerous radioactivity, and (2) specific, laborious degradative schemes had to be devised for the tagged compound to ascertain the position of the labeled atom. The advent of NMR has removed these problems because (1) NMR-active nuclei typically are not radioactive and (2) NMR can detect the site of labeling by simpler, nondestructive means.

For the ^{13}C nucleus, the typical approach to the study of biosynthetic pathways is to feed specifically labeled ^{13}C precursors to organisms, recover the metabolites of interest, and study these products by ^{13}C NMR. Observation in the NMR of enhanced signals then directs one to the positions to which the labeled carbons have migrated. These labeled precursors have commonly included [1-^{13}C]acetate or [2-^{13}C]acetate,[3–5] but a variety of more complex molecules also has been utilized.[5]

Depending upon the level of isotopic enrichment of ^{13}C in the precursors and upon the degree to which the organism incorporates the labeled ^{13}C atoms, the intensity of the ^{13}C signals in the final metabolites may fall in a range from the threshold of 1% natural abundance ^{13}C to levels approaching 100%. The analysis is facilitated if the isotopic enrichment in the final products is at an intermediate value because of carbon–carbon splittings at higher concentrations. If the enrichment in the compounds is 90%, for example, then the probability of two simultaneous ^{13}C atoms in a molecule would be 90% × 90% = 81% and carbon satellites arising from C—C coupling would complicate the situation. If, however, the enrichment is 20–50%, then the probability of two simultaneous carbon atoms is in the range of 4–25% (20% × 20% = 4%; 50% × 50% = 25%) and analysis of the center bonds is more manageable.

Some investigators, alert to the additional structural information available from C—C couplings, have attempted to raise isotopic enrichment levels in the final metabolites so that J_{CC} values could be observed. Originally, only the directly bonded couplings were studied, but subsequently it was realized that long-range couplings could give yet more information regarding the metabolic pathways in the organism. A major contribution to the elucidation of biosynthetic pathways afforded by C—C couplings was that groups of atoms could be followed in the total mechanism. A common technique employed in these studies was to feed a mixture of highly concentrated, multiply labeled precursor and natural precursor, such as $^{13}CH_3$—$^{13}CO_2H$ and CH_3CO_2H. In the final product, the probability of C—C coupling would be much higher if an acetate group were left intact. Indeed, the very point of the experiment would be to establish that the two-carbon group was preserved, cleaved, or rearranged. To illustrate, compounds 1–4 show the statistical distribution if a 1:1 ratio of

$$\underset{1}{CH_3\overset{\overset{O}{\|}}{C}CH_2\overset{\overset{O}{\|}}{C}OH}$$

$$\underset{2}{\underset{*}{CH_3}\underset{*}{\overset{\overset{O}{\|}}{C}}\underset{*}{CH_2}\underset{*}{\overset{\overset{O}{\|}}{C}}OH}$$

$$\underset{3}{\underset{*}{CH_3}\underset{*}{\overset{\overset{O}{\|}}{C}}CH_2\overset{\overset{O}{\|}}{C}OH}$$

$$\underset{4}{CH_3\underset{*}{\overset{\overset{O}{\|}}{C}}CH_2\underset{*}{\overset{\overset{O}{\|}}{C}}OH}$$

*CH$_3$—*CO$_2$H and CH$_3$CO$_2$H is condensed to produce acetoacetic acid. In this example, it is assumed that the level of isotopic enrichment in *CH$_3$*CO$_2$H is 100%. The ^{13}C NMR spectrum of the mixture of 1–4 would show several signals arising from noncoupled and coupled carbon nuclei, and the labeled carbons would be observed above the normal level of natural abundance ^{13}C nuclei. The terminal methyl group, for example, would be seen in 2 and 3. In compound 2 this methyl would be split by the other three carbons to reflect $^1J_{CC}$, $^2J_{CC}$, and $^3J_{CC}$. In compound 3 only the splitting $^1J_{CC}$ would be observed. Hence, $^1J_{CC}$ occurs always, and $^2J_{CC}$ and $^3J_{CC}$ occur 50% of the time. The fact that $^1J_{CC}$ occurs always demonstrates that the terminal acetate group is intact. If acetic acid had experienced carbon–carbon bond cleavage before condensation (admittedly an absurd proposition chemically) then the ^{13}C label would be randomly distributed throughout the final product and the probability of observing $^1J_{CC}$, $^2J_{CC}$, and $^3J_{CC}$ would all be equal. In practice, the ratio of *CH$_3$*CO$_2$H to CH$_3$CO$_2$H should be less than 1:1, perhaps 1:3, in order to reduce the chance of these less probable couplings. It is true that the chance of the more probable coupling is also reduced, but to not nearly so great an extent. The final result is that for a molecule such as acetoacetic acid synthesized from a 1:3 ratio of *CH$_3$*CO$_2$H and CH$_3$CO$_2$H, the terminal methyl signal would appear as a center bond flanked by the ^1J doublet. In actual in vivo experiments the final product is generally diluted with natural compound incorporated from sources other than the material fed to the organism. Accordingly, the relative intensity of the center bond may be larger than that of the flanking split signals. A quantitative treatment of this method has been developed and can be applied to a variety of metabolic types.[6]

An application of this method is illustrated by a study of the pyrone metabolite 7, of the mold *Aspergillus melleus*, in which an acetate unit was shown to undergo intramolecular rearrangement.[7] This unusual biosynthetic pathway was established by the use of a geminal carbon–carbon coupling. The biosynthetic mechanism involves the rearrangement of the pentaketide 5 to give 6, which with decarboxylation gives the product 7. The Favorskii-type rearrangement depicted in 5→6 was established by the observation of a geminal coupling (6.2 Hz) between the marked carbons in 7. All other observed couplings were large and obviously $^1J_{CC}$ values (see 8). These $^1J_{CC}$ values demonstrated intact acetate units, whereas the ^2J value reflected the fact that these two carbons originally belonged to the same acetate group (see **9**).

A quite similar rearrangement was established for vulgamycin (**10**) by noting a geminal C—C coupling as indicated.[8] Again, in this experiment doubly labeled acetate was used, diluted with natural acetate. The geminal coupling of 3 Hz

showed that the two carbons originated from the same acetate molecule and accordingly a Favorskii-type rearrangement was occurring from a biosynthetic intermediate, as shown in **11 → 10**.

The subtlety and effectiveness of using long-range carbon–carbon couplings in the elucidation of biosynthetic mechanisms is particularly demonstrated in studies of porphyrin derivatives.[9] In these studies doubly labeled porphobilinogen (**13**) (synthesized enzymatically from 5-amino[5-^{13}C]levulinic acid, **12**) was converted to uroporphyrinogen-I ester (**14**) and protoporphyrin-IX ester (**15**) by two independent enzymatic reactions. Undiluted porphobilinogen was used to

16   **17**   **18**

prepare **14**. In the ^{13}C NMR of **14**, the four equivalent meso carbon signals (marked with arrows) appeared as a doublet of doublets with $^{1}J = 72$ Hz and $^{3}J = 5$ Hz and demonstrated four equivalent indole units (see **16**) with identical labeling patterns. In **15**, four nonequivalent meso carbons exist and have different chemical shifts. Labeled **15** was prepared from diluted porphobilinogen and consequently couplings would be observed only between labeled carbons that originally resided in the original porphobilinogen **13**. Therefore, each of the meso carbons could be split only into a doublet. The observed doublets are shown in **15**. Three of the doublets had splittings of 5 Hz, as expected in three-bonded couplings as shown in **17**. However, one of the doublets was large (72 Hz), clearly indicative of a one-bonded coupling (see **18**). The conclusion is inescapable—this fourth ring in **15** arises via an intramolecular rearrangement of the porphobilinogen precursor (**13**). Why the incorporation of porphobilinogen (**13**) into **15** involves the integrity of three molecules and the rearrangement of a fourth is a mystery—but the coupling patterns prove this is the case.

Carbon-13 Chemical Shift Assignments

In all biosynthetic studies involving ^{13}C NMR, correct ^{13}C chemical shift assignments manifestly are imperative. Sometimes, however, the complex nature of the ^{13}C NMR pattern renders making these assignments difficult. In such situations additional tools for shift assignments are welcomed, and C—C couplings can sometimes assist in this regard. As an example, the chemical shifts of the aromatic carbons of phenylalanine (**19**) were established by $^{2}J_{CC}$ and $^{3}J_{CC}$

19

values.[10] The ambiguity of the ^{13}C assignments of C-2' and C-3' in **19** was immediately removed with the observation of the J values of 2.5 and 3.5 Hz, respectively. Principles discussed in Chapter 4 (α-Labeled Benzene Compounds) make clear that the larger J value must be the one that is three bonded.

It is possible that coupling patterns can aid simultaneously in elucidating biosynthetic pathways and in making chemical shift assignments. In a study of the biosynthesis of scytalone (20) a $^2J_{CC}$ value of 9 Hz in the multiply labeled compound was observed and helped to remove an ambiguity in ^{13}C shift

20

assignments in the cyclohexanone ring.[11] The large observed coupling is clearly consistent with a geminal C—C coupling across a carbonyl.

Several ambiguities existed in a biosynthetic study with aflatoxin B_1 (21). Chemical shift assignments had been made previously, but a study of multiply

21   **22**

labeled 21 showed that several assignments had been in error.[12] Aflatoxin B_1 produced from $CH_3{*}CO_2H$ gave a labeling pattern shown in 22. Resonances 13 and 15 were reversed because (the correctly assigned) C-15 showed one geminal coupling (to C-13), whereas C-13 exhibited two such couplings (to C-15 and to C-10). Resonances 1 and 6 were also reversed when it was noted that (correctly assigned) C-6 experienced a coupling of 11.4 Hz with C-3. This value of 11.4 Hz is in excellent agreement with the value anticipated for dual-path cyclopentenones (see The Additivity of Carbon–Carbon Coupling Constants, Chapter 5). This example of aflatoxin B_1 clearly points out that previous chemical shift assignments for complex molecules should be regarded critically. Accordingly, it is dangerous to rely heavily on chemical shift assignments based solely on the comparison of chemical shifts of model compounds. As well stated by the authors,[12] "Ideally, the combination of chemical shift and coupling constants should be used for absolute identification." In a similar example ^{13}C chemical shift assignments were confirmed in cholesterol labeled with six ^{13}C nuclei (23).[13] In a biosynthetic study the C-12 signal was split into a doublet of doublets, $^1J = 33$ Hz (to C-11) and $^3J = 4$ Hz (to C-16). This latter coupling is fully in accord with the trans arrangement of the C—C—C—C linkage in question.

Another example where chemical shift assignments were reversed involves

23

monosaccharides.[14] Extensive C—C coupling allowed unambiguous assignments for all carbons. More discussion regarding this topic is deferred to the next section.

It is possible to extend the use of C—C couplings one step further. Accordingly, C—C couplings can be used not only to help elucidate biosynthetic pathways and to aid in chemical shift assignments, but also to assist in the structure elucidation of the metabolite itself. For example, in a study of ochrephilone (**24**), obtained from a penicillin mold, various labeling studies were conducted with *CH_3CO_2H, CH_3*CO_2H, and *CH_3*CO_2H to gain biosynthetic information and even to help elucidate the structure of the metabolite.[15] In this structure determination $^1J_{CC}$ value were utilized extensively, but geminal couplings across the carbonyl were also quite helpful in assigning geometry about the relevant carbons.

24

Saccharides

Table 6-1 reports J_{CC} values for various monosaccharides, including derivatives of glucose (**25–28**), mannose (**29–32**), galactose (**33, 34**), fucose (**35, 36**), acetate derivatives (**37–40**), and amino sugars (**41–44**). Chemical shift assignments were based, in part, on observed coupling constants and it was noted[14] that previous predictive tools for making shift assignments had been erroneous.

Geminal C—C couplings in the monosaccharides show a remarkable dependence upon the orientation of the anomeric substituent. When the substituent is α (**50**), the ^2J value to C-3 is small and to C-5 is substantial (~ 2 Hz). By contrast, for the β anomer (**51**), the coupling is large to C-3 (~ 4 Hz) and is small to C-5. These findings can be rationalized in terms of the principles

50 51

outlined earlier. If $^2J_{CC}$ couplings have a similar dependence of sign upon orientation of a terminal electronegative substituent as have $^2J_{CH}$ couplings (see Geminal J Values in Conformational Analysis, Chapter 2), then an anti OH group will have a positive contribution to $^2J_{CC}$. If it is then assumed that the geminal coupling to C-5 in **50** is negative, then it is reasonable that this coupling to C-5 in **51** is zero, since the —OH group in **51** is anti to the coupling path. That the $^2J_{CC}$ coupling to C-5 in **50** is negative is well in accord with previous principles: geminal coupling across oxygen is negative (see Symmetrical Double-Labeling Method, Chapter 5) and the gauche OH substituent in **50** should have little effect upon the sign. Meanwhile, this gauche —OH substituent should also have minimal effect upon the terminal coupling to C-3, and the observed near-zero coupling is consistent with observations based upon other corresponding aliphatic couplings (Table 3-1). It follows that the coupling to C-3 in **51** should be positive. The very large value of ^2J to C-3 in **51** may result from the presence of a second anti OH group, substituted on C-3. Corroboration of this speculation must await sign determinations. In the meantime, the empirical trends depicted in **50** and **51** should prove useful in conformational studies of monosaccharides.

Regarding the vicinal couplings, ^3J values to C-4 are small, a result that is not unexpected because the dihedral angle is near 60° (see **52** and **53**). The vicinal coupling to C-6 is larger because the dihedral angle is near 180°. Of the two anomers, the C-6 coupling in the β anomer (**53**) is larger, as expected from previous studies (see Table 3-9), where it was seen that an anti electronegative substituent increases $^3J_{CC}$.

52 53

Table 6-2 lists some carbon–carbon couplings for five disaccharides, each with two pyranose units. The geminal and vicinal couplings with one pyranose ring closely follow the trends for monosaccharides as portrayed in **50–53** and accordingly attest to the existence of undistorted pyranose units. The new type of data in disaccharides is, of course, a result of the existence of couplings

TABLE 6-1. CARBON–CARBON COUPLINGS (AND CARBON-13 CHEMICAL SHIFTS) OF MONOSACCHARIDES

| Compound | J_{CC}(Hz) to labeled carbon (δ_{C-13}, ppm) of carbon no.: | | | | | | | |
	1	2	3	4	5	6	OCH₃	Reference
25 α-D-glucopyranose	(93.6)	46.0 (73.2)	~0 (74.5)	~0 (71.4)	1.8 (73.0)	a (62.3)		14
25a α-D-glucopyranose	3.2		3.7		43.1			16
26 β-D-glucopyranose	(97.4)	46.0 (75.9)	3.5 (77.5)	~0 (71.3)	~0 (77.4)	a (62.5)		14
26a β-D-glucopyranose	3.8		4.4		43.3			16
27 methyl-α-D-glucopyranose	(100.6)	46.4 (72.7)	~0 (74.7)	~0 (71.2)	1.7 (73.0)	3.2 (62.2)	(56.5)	14
28 methyl-β-D-glucopyranose	(104.6)	46.8 (74.6)	4.1 (77.4)	~0 (71.2)	~0 (77.3)	4.3 (62.4)	(58.5)	14

TABLE 6-1. (continued)

Compound	J_{CC}(Hz) to labeled carbon (δ_{C-13}, ppm) of carbon no.:							
	1	2	3	4	5	6	OCH$_3$	Reference
29 α-D-mannopyranose	(95.5)	46.8 (72.3)	~0 (71.9)	~0 (68.5)	1.7 (73.9)	a (62.6)		14
30 β-D-mannopyranose	(95.2)	42.4 (72.8)	4.3 (74.8)	~0 (68.3)	~0 (77.6)	a (62.6)		14
31 methyl-α-D-mannopyranose	(102.2)	47.0 (71.4)	~0 (72.1)	~0 (68.3)	2.3 (73.9)	3.0 (62.5)	(56.1)	14
32 methyl-β-D-mannopyranose	(102.3)	43.8 (71.7)	3.4 (74.5)	~0 (68.4)	~0 (77.6)	4.0 (62.6)		14
33 α-D-galactopyranose	(93.8)	46.6 (70.0)	~0 (70.8)	~0 (70.9)	2.1 (72.0)	a (62.8)		14
34 β-D-galactopyranose	(98.0)	46.0 (73.6)	3.7 (74.4)	~0 (70.4)	~0 (76.6)	a (62.6)		14

TABLE 6-1. (continued)

| Compound | J_{CC}(Hz) to labeled carbon ($\delta_{C\text{-}13}$, ppm) of carbon no.: | | | | | | | |
	1	2	3	4	5	6	OCH$_3$	Reference
35 α-L-fucopyranose	(93.8)	45.8 (69.8)	~0 (71.0)	~0 (73.5)	2.3 (67.8)	3.5 (17.2)		14
36 β-L-fucopyranose	(97.8)	46.0 (73.4)	4.1 (74.6)	~0 (73.1)	~0 (72.3)	3.5 (17.2)		14
37 methyl α-D-glucose tetraacetate	47.0	0			2.0	3.5	2.0	16
38 methyl β-D-glucose tetraacetate	48.8	4.6			0	3.1		16
39 α-D-glucose pentaacetate	47.7	0						16
39a α-D-glucose pentaacetate	3.8		3.3		45.1			16

TABLE 6-1. (continued)

Compound	J_{CC}(Hz) to labeled carbon (δ_{C-13}, ppm) of carbon no.:							
	1	2	3	4	5	6	OCH$_3$	Reference
40 AcO—CH$_2$OAc / O / AcO— *—OAc AcO (β-D-glucose pentaacetate)		49.4	5.2		0	4.3		16
40a AcO—*CH$_2$OAc / O / AcO— —OAc AcO (β-D-glucose pentaacetate)	4.4			3.8	44.5			16
41 HO—CH$_2$OH / O / HO— * H$_2$N OH (α-D-glucosamine-DCl)		44.3	~0	~0	1.5	a		17
42 HO—CH$_2$OH / O / HO— *—OH H$_2$N (β-D-glucosamine-DCl)		43.8	2.4	~0	~0	a		17
43 HO—HOCH$_2$ NH$_2$ / O / HO— * OH (α-D-mannosamine-DCl)		44.7	~0	~0	1.7	a		17
44 HO—HOCH$_2$ NH$_2$ / O / HO— —OH * (β-D-mannosamine-DCl)		41.3	1.8	~0	~0	a		17

aThese couplings were reported as observed but not measured.

TABLE 6-2. CARBON–CARBON COUPLINGS IN DISACCHARIDES[a]

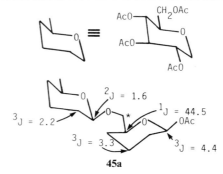

45a

Gentiobiose octaacetate

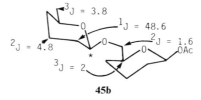

45b

Gentiobiose octaacetate

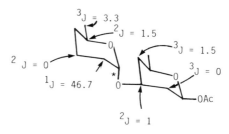

46

Nigerose octaacetate

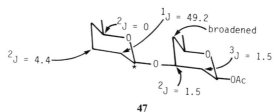

47

Laminarabinose octaacetate

TABLE 6-2. (continued)

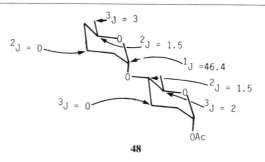

48

Maltose octaacetate

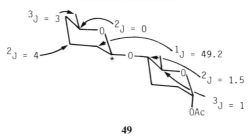

49

Cellobiose octaacetate OAc

[a]Reproduced with permission.[16]

between pyranose units through the anomeric oxygen atom. Geminal couplings of this type lie consistently in the range 1–1.6 Hz and probably reflect the small negative couplings through an ether linkage (see Symmetrical Double-Labeling Method, Chapter 5). The vicinal couplings are more variable (0–2.2 Hz) and no doubt reflect different rotational geometries between the two pyranose units. A detailed discussion of conformational analysis of these disaccharides utilizing Karplus-like relationships has appeared.[16] It will suffice here to note that the standard orientations of the pyranose units as portrayed in **45–49** of Table 6-2 cannot be correct—for example, in **54** (taken from **48**, Table 6-2) the two

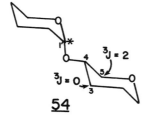

54

different vicinal couplings (0 and 2 Hz) indicate twisting about the glycosidic linkage to allow the 1′–0–4–5 angle to approach planarity while the 1′–0–4–3 angle is more nearly orthogonal.

Amino Acids and Peptides

Table 6-3 includes various vicinal C—C couplings for amino acids, obtained by the ^{13}C NMR analysis of compounds enriched at various positions. A cursory inspection of the C-1–C-4 vicinal couplings (see **55**) shows a rather wide

$$\underset{4}{C}-\underset{3}{C}-\underset{2}{\overset{\overset{\displaystyle NH_2}{|}}{C}H}-\underset{1}{CO_2H}$$

55

range of J values that seems to indicate varying conformations for different amino acids. Confusing the issue, however, is the very wide range of pH which may in itself change the conformation of the amino acids. Also notable in Table 6-3 are the two $^3J_{CC}$ values in valine, reflecting the diasterotopic nature of the two γ-methyl groups (see **56**).

TABLE 6-3. VICINAL CARBON–CARBON COUPLING CONSTANTS OF AMINO ACIDS AND PEPTIDES

Compound	$^3J_{1\text{-}4}$	$^3J_{2\text{-}5}$	$^3J_{3\text{-}6}$	pH
Valine[a]	2,4,1.0			11.2
Threonine[a]	2.3			0.7
Aspartic acid[a]	3.3			6.5
Glutamic acid[a]	1.5	3.4		0.9
Pyroglutamic acid[a]	1.1	7.5		1.0
Lysine[a]	~0.6	4.3	4.7	7.0
Arginine[a]	~0.5	4.8	~0.5	7.6
Phenylalanine[b]	2.5			Organic solvent
Proline[c]	1.5			
Gly-Pro[c,d]	<0.5	4.4		–
Gly-Pro-Gly[c,d]	<0.5	4.3		
TRF[c,d,e]	<0.5	3.9		

[a] Reference 19.
[b] Reference 10.
[c] Reference 20.
[d] J in proline residue.
[e] TRF = thyrotropin releasing factor, <Glu-His-Pro-NH$_2$.

$$H_3C-CH(CH_3)-CH-CH(NH_2)-CO_2H \text{*}$$

56

To dramatize the difficulty of comparing the $^3J_{CC}$ values of different amino acids at different pH values, the $^3J_{CC}$ values of individual amino acids have been well demonstrated to vary considerably over the pH scale of ca. 0–13.[18] In this study, aspartic and glutamic acids were shown to exhibit increasing ^3J values as the pH was increased, when either J_{1-4} or J_{2-5} was considered. For example, in aspartic acid (**57**) the experimental $^3J_{1-4}$ values increase from 2.6 to 4.0 Hz over

$$J_{1,4}$$
$$NH_2$$
$$HO_2C\text{-}CH_2\text{-}CH\text{-}CO_2H$$

57

the pH range of 0–12. This observation is indicative of the presence of increasing amounts of anti population, from the Karplus-like relationship discussed in Chapter 3. It is in fact known from other studies that the anti population of aspartic acid increases with increasing pH. Interestingly, as the pH is increased over the range of 0–12, the plotted $^3J_{CC}$ values show plateaus reflecting the various ionization states of aspartic acid. At low pH gauche conformations (**58**) are more favored, and at high pH the anti conformation (**59**) is more favored with its carboxylate anionic charges. An additional complication, of course, is

58   **59**

that ^3J of a carboxylic acid for a given conformation would be expected to vary somewhat with changing ionization state of the carboxyl groups.

The J_{2-5} values have been measured for glutamic acid (**60**) as a function of

$$J_{2,5}$$
$$NH_2$$
$$HO_2C\text{-}CH_2CH_2\text{-}CH\text{-}CO_2H$$

60

pH.[18] The dependence of J_{2-5} upon pH is similar to that of J_{1-4} for aspartic acid, i.e., higher J values at high pH. The entire range is 3.0–4.4 Hz over pH 0–13. A secondary minimum occurs at pH 7, and obviously there is substantial interplay between the conformational and electronic factors.

A few data exist for peptides (Table 6-3) in which ^{13}C-enriched proline residues have been incorporated. The values for J_{1-4} suggest different conformations for the pyrrolidine ring in free proline ($J = 1.5$ Hz) and in various peptides ($J < 0.5$ Hz).[20] Free proline is known to rapidly interconvert between the two puckered forms 61 and 62, which would lead to an overall average value of a

large J (from 61) and a small J (from 62). The small value actually observed for the peptides (< 0.5 Hz) suggests that conformation 62 is preferred in these compounds. It has been suggested[20] that adopting such a conformation leads to maximum hydrogen bonding between residues in the peptide.

Heterocyclic Compounds

In 63–68 are listed the long-range C—C couplings in barbitals,[21] dilantin,[21] and pyroglutamic acid.[19] The effect of a heteroatom can be quite unusual, and some couplings are difficult to understand. In the barbitals ^2J values from the

67   **68**   **69**

carbonyl are ca. 2–3 Hz, whether or not through a heteroatom (3.13 and 2.22 Hz in **63**, 3.16 Hz in **65**). The ^2J values in the barbitals from an sp^3 carbon appear normal (< 1.0 to an aliphatic carbon and 2.91 Hz to an aromatic carbon in **64**). In the barbitals the transannular vicinal coupling is ca. 1 Hz (1.10 Hz in **64**, 1.07 Hz in **66**). The vicinal coupling to the aromatic ring in **64** is less than 1 Hz; this small value is quite perplexing.

In dilantin (**67**) two dual-path couplings are reported, and it is puzzling why their values are so different. In pyroglutamic acid (**68**) the reported dual-path coupling is 7.5 Hz. A model compound for **67** and **68** might be **69** (taken from Table 4-11). The value of 6.57 Hz in **69** is quite in accord with $J = 8.10$ Hz in **67** and $J = 7.5$ Hz in **68**, but the $J = 6.87$ Hz in **69** is quite dissimilar to $J = 1.43$ in **67**. Apparently the three-bonded path C-2—N-3—C-4—C-5 in **67** is near zero, and the observed coupling to C-5 reflects only the ^2J value resulting from the pathway C-2—N-1—C-5. In fact, it was noted in Chapter 5 that a vicinal coupling with an internal carbonyl can be quite small.

70

In another study,[22] a vicinal coupling in histidine (see **70**) was studied as a function of pH. At low pH values a low ^3J value was observed (2.1 Hz), whereas at high pH values a larger ^3J value of 4.3 Hz was seen. Over the pH range of 1–11 two plateaus were observed at either extreme of the plot and at pH ~ 7 a dramatic change in ^3J occurred. It is apparent that the protonated species effects a less efficient transmission of coupling, whereas at a high pH the vicinal coupling value appears more nearly normal.

The small amount of data regarding heterocycles makes it difficult to make generalizations, and until more studies have been completed uncertainties such as those discussed above will remain. In the meantime, model compounds taken from studies discussed in Chapters 3–5 must be applied with caution.

References

1. Urey, H. C.; Brickwedde, F. G.; Murphy, G. M. *Nature (London)*, **133**, 173 (1934).
2. Murray, A. III; Williams, D. L. "Organic Synthesis with Isotopes," Parts 1 and 2; Wiley-Interscience: New York, 1958.
3. Cushley, R. J.; Anderson, D. R.; Lipsky, S. R.; Sykes, R. J.; Wasserman, H. H. *J. Am. Chem. Soc.*, **93**, 6284 (1971).
4. Polonsky, J.; Baskevitch, Z.; Cagnoli-Bellavita, N.; Ceccherelli, P.; Buckwalter, B. L.; Wenkert, E. *J. Am. Chem. Soc.*, **94**, 4369 (1972).
5. Omura, S.; Takeshima, H.; Nakagawa, A.; Miyazawa, J.; Piriou, F.; Lukacs, G. *Biochemistry*, **16**, 2860 (1977).
6. Gagnaire, D.; Taravel, F. R. *J. Am. Chem. Soc.*, **101**, 1625 (1979).
7. Simpson, T. J. *Tetrahedron Lett.*, 4693 (1975).
8. Seto, H.; Sato, T.; Urano, S.; Uzawa, J.; Yonehara, H. *Tetrahedron Lett.*, 4367 (1976).
9. Battersby, A. R.; Hunt, E.; McDonald, E. *Chem. Commun.*, 442 (1973).
10. Leete, E.; Kowanko, N.; Newmark, R. A. *J. Am. Chem. Soc.*, **97**, 6826 (1975).
11. Sankawa, U.; Shimada, H.; Sato, T.; Kinoshita, T.; Yamasaki, K. *Tetrahedron Lett.*, 483 (1977).
12. Hsieh, D. P. H.; Seiber, J. N.; Reece, C. A.; Fitzell, D. L.; Yang, S. L.; Dalezios, J. I.; LaMar, G. N.; Budd, D. L.; Motell, E. *Tetrahedron*, **31**, 661 (1975).
13. Popják, G.; Edmond, J.; Anet, F. A. L.; Easton, N. R. Jr. *J. Am. Chem. Soc.*, **99**, 931 (1977).
14. Walker, T. E.; London, R. E.; Whaley, T. W.; Barker, R.; Matwiyoff, N. A. *J. Am. Chem. Soc.*, **98**, 5807 (1976).
15. Seto, H.; Tanabe, M. *Tetrahedron Lett.*, 651 (1974).
16. Gagnaire, D. Y.; Nardin, R.; Taravel, F. R.; Vignon, M. R. *Nouv. J. Chim.*, **1**, 423 (1977).
17. Walker, T. E.; London, R. E.; Barker, R.; Matwiyoff, N. A. *Carbohydr. Res.*, **60**, 9 (1978).
18. London, R. E.; Walker, T. E.; Kollman, V. H.; Matwiyoff, N. A. *J. Am. Chem. Soc.*, **100**, 3723 (1978).
19. Tran-Dinh, S.; Fermandjian, S.; Sala, E.; Mermet-Bouvier, R.; Fromageot, P. *J. Am. Chem. Soc.*, **97**, 1267 (1975).
20. Haar, W.; Fermandjian, S.; Vicar, J.; Blaha, K.; Fromageot, P. *Proc. Natl. Acad. Sci. U.S.*, **72**, 4948 (1975).
21. Long, R. C. Jr.; Goldstein, J. H. *J. Magn. Reson.*, **16**, 228 (1974).
22. London, R. E. *Chem. Commun.*, 1070 (1978).

General Index

Pages marked with an asterisk (*) are tables.

Pages marked with an asterisk (*) are tables.

Pages marked with an asterisk (*) are tables.

Pages marked with an asterisk (*) are tables.

Pages marked with an asterisk (*) are tables.

Pages marked with an asterisk (*) are tables.

Pages marked with an asterisk (*) are tables.

Pages marked with an asterisk (*) are tables.

Pages marked with an asterisk (*) are tables.

Compound Type Index

Pages marked with an asterisk (*) are tables.

ALIPHATIC COMPOUNDS: MONOCYCLIC

ALIPHATIC COMPOUNDS: POLYCYCLIC

Pages marked with an asterisk (*) are tables.

Pages marked with an asterisk (*) are tables.

π-SYSTEMS: GENERAL

OLEFINS: GENERAL

OLEFINS: ACYCLIC

Pages marked with an asterisk (*) are tables.

Pages marked with an asterisk (*) are tables.

OLEFINS: CYCLIC

ACETYLENES

Pages marked with an asterisk (*) are tables.

Pages marked with an asterisk (*) are tables.

AROMATICS: POLYNUCLEAR

Pages marked with an asterisk (*) are tables.

AROMATICS: OTHER

HETEROCYCLIC COMPOUNDS

Pages marked with an asterisk (*) are tables.

BIOCHEMICAL SYSTEMS

MISCELLANEOUS

Pages marked with an asterisk (*) are tables.